SpringerBriefs in Electrical and Computer Engineering

More information about this series at http://www.springer.com/series/10059

Sunil Kumar Kopparapu

Non-Linguistic Analysis of Call Center Conversations

 Springer

Sunil Kumar Kopparapu
TCS Innovation Labs—Mumbai
Thane (West)
Maharashtra
India

ISSN 2191-8112   ISSN 2191-8120   (electronic)
ISBN 978-3-319-00896-7   ISBN 978-3-319-00897-4   (eBook)
DOI 10.1007/978-3-319-00897-4

Library of Congress Control Number: 2014944343

Springer Cham Heidelberg New York Dordrecht London

Printed on acid-free paper

Springer is part of Springer Science+Business Media (www.springer.com)

Nana, Namita and Sumit; with whom I try spend all the time when away from work.

Remembering Amma and Avva.

Preface

Voice-based call centers or business process outsourcing units generate huge amounts of speech data everyday during their day-to-day operations. Large and diverse types of information are hidden in these natural language conversations, which is begging to be exploited. The whole area of voice analytics deals with the aspect of deriving *usable* information from the audio data.

Conventionally, speech data converted to text followed by natural language text processing has been used to derive analytics. However, this traditional process of analyzing audio conversations has several major limitations. On one hand, the fact remains that conversion of natural language spoken conversation into text is still maturing even for languages which are well researched and rich in language resources, like English, which hampers the process of analyzing call center audio conversations. On the other hand, understanding aspects of audio conversations by text analysis is not comprehensive. How does one distinguish a /thank you/ spoken in jest and sarcasm versus /thank you/ spoken with gratitude by analyzing just the text "*Thank You*"? Further from a call center perspective, many a times the audio conversation needs to be analyzed with the simple requirement to spot an abnormal call from a normal call in which case the process of speech to text conversion would be an overkill.

In this short monograph, we will dwell on how non-linguistics features associated with spoken conversation can be used to infer information embedded in the call conversation. While the use of non-linguistic analysis can give insight into important aspects of how the conversation happened without worrying about what is the actual linguistic content of the conversation. Additionally, non-linguistic analysis eliminates the need to adopt a not-so-reliable speech to text conversion process, which gives us the flexibility of being able to analyze conversation with little dependency on the content and language of conversation.

This monograph is divided into six chapters. Chapter 1 gives an overview of the process of spoken articulation from an idea or a thought and tries to bring out the different facets of information that is embedded in spoken speech like the owner of the speech, the content of the speech and style of speech. The second part shifts to give an idea of why people still use telephone channel to sort out their problems even under the scenario of several channels being available to them. This motivates the need for analyzing call center conversation and gives some insight into how the enterprise can benefit by deep diving into audio conversation data.

Chapter 2 covers the general process adopted for voice analytics after giving an idea of how a typical audio conversation at a call center is initiated and recorded for analysis. We also cover the entire process to convert the recorded audio conversation to make the audio data suitable for analysis. There is a section on music voice separation, speaker separation, and speech to text conversion. We portray the challenges involved in the process of speech to text.

Chapter 3 is brief and dwells into the constituent of a typical call conversation and highlights the drawback of a manual analysis which points toward the need for an automated analysis of the call conversation.

Chapter 4 talks of the non-linguistic speech processing of call center audio conversation. We specifically speak in detail of two speech features, the speaking rate, and the emotion in speech.

Chapter 5 is essentially a case study which gives details of how non-linguistic features can be used to distinguish a normal call from an abnormal audio conversation based on some analysis of some real call center audio conversation recordings and we conclude in Chap. 6. All the chapters are amply supported by additional reading material in the form of Appendices and references.

Call Center Voice Analytics is an important area of work and constantly evolving, especially with (a) significant growth in services industry and (b) people sticking to telephone channel and not demonstrating any sign of shifting to other channels for customer care interaction. We believe this monograph will be of use to practicing engineer as well as researchers working on speech signal processing. You should be able to follow updates at https://sites.google.com/site/nlccanalytics/.

Thane (West), Maharashtra, June 2014   Sunil Kumar Kopparapu
India (Loc: 72.977265, 19.225129)

Acknowledgments

Several people and things in very many different ways have contributed to this monograph. Acknowledge, from the bottom of my heart, in no particular order.

- Meghna,
- Imran,
- LaTeX,
- GoBiYa!,
- Springer,
- WorldComp 2012,
- Sumit,
- Namita,
- K S Rao,
- Prof PVS Rao,
- Dr. Pande,
- 5G4, Yantra Park,
- Speech Lab,
- Tata Consultancy Services,
- TCS Innovation Labs—Mumbai,
- xfig,
- Linux.

The work reported here is based on discussion and interaction with the team at TCS Innovation Labs—Mumbai. I have liberally made use of the writeups, documents, presentations that we generated together. This is duly acknowledged.

Special mention of thanks to Krutz for initiating this monograph and Rebecca for patiently following up even as I kept postponing the final draft. If you are reading this monograph, it is because of Rebecca.

Lastly, I take complete responsibility for any defects in this monograph, be it typo, grammar, or content.

Contents

Chapter 1
Overview

1.1 Introduction

Speech is the most natural form of communication that has been in existence for a very long time now. Intelligence, which distinguishes homosapiens from the rest of the animal species can be attributed to the development of language as a tool and speech as a mode of communication.

> Communication among species was necessitated by the basic need for survival and, in a way, to keep them from extinction.

Ability to communicate on one hand helped in finding food and avoiding predators, while on the other hand, it helped in finding a mate which helped in reproduction, perpetuation of their own species. In higher species, a combination of sounds and body gestures formed the mode of communication. This combination gave the higher species the ability to communicate emotions in terms of anger, love, concern.

In humans, the evolution of speech has been that of the higher species, namely predominantly sound and gesture based. Conveyance of emotions was used more or less to communicate concepts and the sound and body gesture aspect was predominantly dominated by sound. As an effect, in humans, this led to not only freeing of the limbs and body but also enabled making line of sight communication not crucial. The overall effect of this was that it facilitated *organization of group activity* like hunting and fighting, which helped in gaining dominance over other species.

Communicating concepts, however, required a language. Language can be defined as the systematic creation and usage of a set of symbols, where each symbol is paired with an intended meaning and the meaning is not always absolute but established through social conventions. So in a very crude sense, language is a system of symbols for encoding and decoding information (see Fig. 1.1). The language used by humans can be termed as natural language and is full of intended meanings. The language rules are generally very tricky and complex making it difficult for machines to comprehend the meaning of a natural language text. The whole field of natural language processing (NLP) in the broad area of artificial intelligence dwells on enabling the ability of machines to comprehend natural language spoken or text messages.

© The Author(s) 2015 1
S.K. Kopparapu, *Non-Linguistic Analysis of Call Center Conversations*,
SpringerBriefs in Electrical and Computer Engineering,
DOI 10.1007/978-3-319-00897-4_1

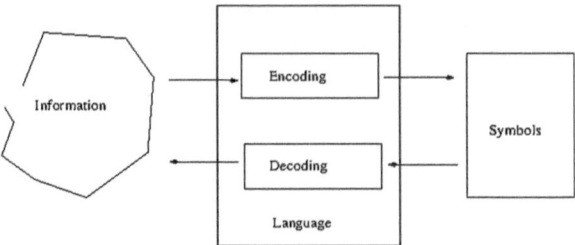

Fig. 1.1 Language helps in mapping information to symbols

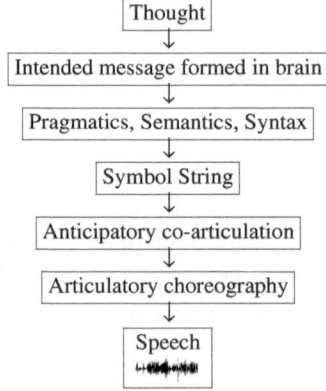

Fig. 1.2 From thought to speech: a top-down view

Speech can be equated to natural spoken language and hence all the complexities associated with the language are transferred to spoken natural language conversation. A simplified (for a comprehensive view see [1]) view of how speech is generated from thought is captured in Fig. 1.2.

The seed of any spoken speech is a thought; the intended message is formed in the brain as a cloud of disconnected objects which are chained together to form a realistic or a practical (pragmatic) connection between the objects. These objects are connected with the meaning of language (semantics) and a grammatical arrangement of words (syntax) to generate a symbol string. These isolated string of symbols are coarticulated to produce speech sound. Coarticulation is the characteristic of a speech sound associated with a particular symbol which is dependent on the preceding symbol (anticipatory coarticulation) [2]. These speech sounds are arranged carefully to produce the final speech. Figure 1.3 shows an example of a complete process of how a thought is processed and converted into speech. As seen in Fig. 1.3, the whole process starts with a thought. In this case, the thought could be to speak or ask

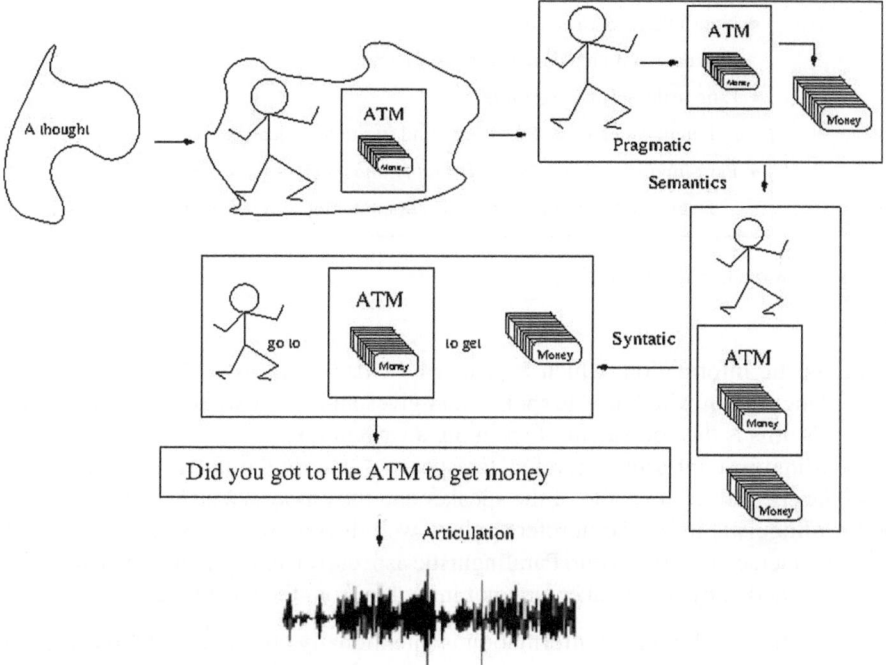

Fig. 1.3 From thought to speech: example

/Did you get the money from an ATM/

or speak

/Sunil got some money from an automated teller machine/.

The *thought* translates into a set of objects which are essentially a person, the concept of money and that of an ATM. The relationship between these objects is established and each of these objects is associated with the language in terms of assigning a symbolic word to each of these objects; in this case, assigning the word automated teller machine, Sunil or "you" to the person object and "money" or "cash" to the currency object. These are stitched together to form "Did you go to the ATM to get cash" or "Sunil got some money from the ATM" as the case might be. These are converted into a symbol string which produces sound; these pieces of sounds are uttered keeping in mind the previous sound while uttering the current sound. Finally, the style, emotion, and situation effects the choreography to produce the resulting speech sound.

> *There are aspects of personal characteristics that come into existence in the spoken speech. Some of which are unintentional like the pitch and the tone of the voice; and some are intentional and purposely controlled by the speaker.*

There are broadly three different types of information embedded in a spoken speech (see Fig. 1.4). They can be broadly classified as:

- Non-linguistic, (*who said it*)
  - gender, emotional states, speaker name
- Linguistic (*what (s)he said*)
  - Language name and what was said (written language)
- Para-linguistic (*how well said, quality; manner, clarity, accent*)
  - Deliberately added by speaker; not inferable from written text.

Fig. 1.4 Information in speech

- Linguistic information, which is primarily what is said. It could be determining the language in which it was spoken and gives an idea of what was said; in other words, this is the equivalent of speaking a written text.
- Non-linguistic information, refers to aspects of who said it in terms of say gender of the speaker or the name of the speaker and the emotional state of the speaker.
- Para-linguistic information refers to how well the text was spoken. It has aspects of manner, clarity, or accent. Paralinguistic aspects related to quality that are deliberately added by the speaker, and not inferable from the written text.

While the paralinguistic information is intentionally controllable by the speaker, the non-linguistic information is intentionally uncontrollable. In this monograph we will call both these intentional and unintentional characteristics introduced into the spoken speech by the speaker as non-linguistic speech features. Equivalently, anything that is not linguistic about spoken speech is termed non-linguistic.

1.2 Call Center Conversations

There has been a tremendous growth in the services sector, which translates to an enterprise staying connected with their customer all the time. Centralized customer contact centers provide the framework for customers to contact the enterprise when a need arises. Typically, a contact center provides several interaction channels to their customers and the voice telephone channel is one of them. In a typical voice-based call center, the customer is speaking to a human agent representing the enterprise. The customer and the agent converse with the aim of resolving the problem that the customer might have. We term this voice conversation between the customer and the agent as call center conversation.

Competition and globalization has led to growth in customer service centers mushrooming in all parts of the globe, meaning customers of one geography are serviced by agents representing the enterprise in another geography. Under these circumstances, the ability of the agents to satisfy the customers who call in to the customer service centers gains importance. Overall, a business measures the effectiveness of the performance of a contact center based on the its ability to satisfy

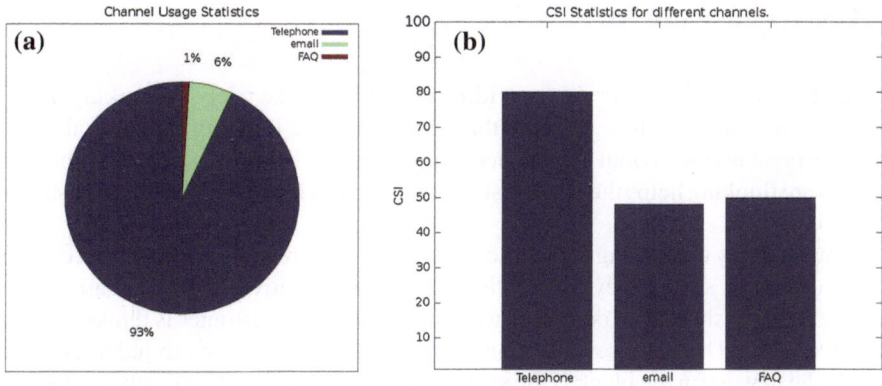

Fig. 1.5 **a** Channel usage statistics; **b** Overall customer satisfaction is 77 on a scale of 100, the customer satisfaction is 80 for telephone, while it is 48 for email and 50 for web FAQ

the actual customer in addition to the economic feasibility of running a contact center. In this monograph, when we mention call center, we mean that it is a voice-based call center.

Customer Satisfaction Index (CSI) is one of the most important metrics used to measure the performance of a voice-based call center. To make it convenient for the customer, enterprises use all the interaction channels that exist, namely telephone, email, web FAQ, SMS. However, studies have shown that the majority of the customers choose the telephone channel to interact for any customer care support. Especially, customers like to speak to the human agent (see Fig. 1.5a). Further, it has been observed that there is higher customer satisfaction for a phone or phone-based interaction (see Fig. 1.5b) compared to other existing interaction channels. While there could be several reasons for this, it is not hard to guess the possible reasons for a higher customer satisfaction in a telephone interaction channel:

(a) Telephone channel is more personal.
(b) Telephone channel enables live and real-time interaction with the agent at the other end.
(c) Speech is the most natural mode of communication to sort problems.
(d) It is a synchronous transaction.
(e) Closest to a face-to-face interaction.

Statistics show that the customers' usage of the available interaction channels is highly biased toward voice channel for interaction (see Fig. 1.5). This means that any dissatisfaction, however minute, in customers who use voice for customer care will touch a large number of users and hence there is a need to analyze voice interactions between the customer and the agents in a call center. Analyzing these interactive conversations between the agents and customers can assist in quickly identifying any issues and addressing them quickly before they spiral out of control and induce customer dissatisfaction.

1.3 Need for Analysis

Several customer facing service providing enterprises (like telecom, banking, insurance) need to have a pulse on what their customers are thinking (a) about them as an enterprise and (b) about the services they provide. Knowledge of what the customers are thinking helps the enterprise tweak their services to induce better customer satisfaction.

Among many other things, it is essential for a service rendering enterprise to retain their customers, so that they can continue to stay competitive. Customer satisfaction index or CSI in short is a good measure to judge what the customer is thinking about the enterprise. There are several channels that an enterprise uses to judge the CSI. While the newest on the block is the social networking channel like facebook, twitter, and personal blogs. The channels that are a rich source of information are the points of interaction (PoI) between the customers and the enterprise. While SMS, email, and live chat are some channels of interactions, the majority of the interaction even today is telephone channel through the telephone channel (see Fig. 1.5).

Customer service centers are gaining importance because they have become the information decimating focus point of an enterprise. Monitoring customer–agent interaction serves in decision making in terms of trying to improve CSI. While companies try to optimize by providing customer service centers in areas which make economic sense, they are cautious that the ultimate customer derives the maximum service without hardship because a dissatisfied customer has a direct bearing on the business performance. This motivates enterprises to monitor and analyze speech transaction between the agent and the customer.

Agent-based voice call center are for many enterprises a rich source of information as they handle telephone calls directly from customers of such enterprises. Large call centers handle very large volumes of telephone calls from customers and for a variety of reasons, these enterprises would like to analyze on a day-to-day or real-time basis the audio conversation between the customer and the agent to derive usable information from customer–agent audio telephonic conversations. All the while, the prime focus is still on trying to keep a tab on the CSI, namely

Understand the customer needs and address them quickly and efficiently and thereby enhance customer satisfaction.

However, there are several other dimensions that the analysis of these conversations could unearth. For example, analysis of interaction between the agent and the customer could:

(a) Help streamline business processes and improve overall operational efficiency of the call center itself in addition to reducing the operational costs of running a call center
(b) Help improve call center voice agent expertise by identifying personalized training needed by the agents

(c) Gather market intelligence, like the mention of a competitor name by the customer, embedded in the conversation and help make quick business decisions on products or processes.
(d) Monitor compliance, regulated by law of the land, to maintain adequate standards of behavior and practice, especially in terms of the agents dialog with the customer.

Apart from these, the increasing need to analyze call center voice conversation to derive usable analytics stems from the fact that speech analytics can assist the business performance in several ways [3], for example,

1. Improve First-Call Resolution Rates (FCR)
   Improved FCR implies improving the number of calls from the customers that have been satisfactorily solved without requiring (a) escalation to the next level agent or (b) the need for the customer to call again. This is one of the most watched metrics in all modern call center and is a key contact center performance metric for two reasons: (i) Each escalation of subsequent call increases the cost of providing that solution and (ii) an additional call to solve the same query tends to lower CSI.
   Voice analytics can help identify the *actual* cause for a poor FCR by analyzing call conversations by identify factors that results in repeat calls. The identified problem areas can be ironed out by allowing for implementation of necessary steps in the call center process.
2. Reducing Average Handle Times
   Average handle time (AHT) has a direct impact on the performance of the call center. Average Handle Time is the statistical average time it takes for calls to be handled by an agent. It generally includes the actual interaction time with the customer plus after call work time to make notes of the actions that need to be taken post the call. Optimizing AHT can strike the right balance between the quality of the customer experience and the costs associated with agent's time.
   Analyzing speech audio conversations can be used to not only measure AHT trends but also help identify reasons for its unexpected variations. This analysis can assist to identify where to apply additional training, staffing changes, procedural adjustments, and so on, to effect improvement.
3. Repairing Broken Processes
   Call centers in general have several mandated processes; subsequently, there is a risk of any one of these leading to problems affecting FCR and AHT. Broken process could be faulty handsets used by voice agents to inappropriate procedural adjustments.
   Analysis of speech conversations can determine recurring events; for instance dropped calls, while serving a particular type of call, or calls from a particular region. It can quantify the number of times that a broken process happens, which can in turn highlight underlying process problems, thereby helping build a business plan for new resources.

4. Quality Monitoring of Voice Agent

   Quality of interaction of the agent with the customer can have a significant impact
   on customer satisfaction. Supervisors in call centers are only able to assess about
   2–10 randomly selected calls a month per agent. Clearly, this is far from sufficient
   to monitor the quality of performance of the agent, additionally, there is no consis-
   tency in the types of calls assessed which leads to difficulty in identifying specific
   behaviors associated with certain types of calls that might be negatively impacting
   agent performance. Lack of this information means that targeted coaching of the
   agent to improve success is very limited, as a result agent becomes discouraged,
   ineffective, and at risk of attrition.

   Automatically analyzing speech conversations can help monitor *all* the conversation in the call center. This ability provides a level of insight and understanding
   of 100 % agent performance which is essential for targeted coaching of agents.

5. Reducing Customer Dissatisfaction

   Measuring customer dissatisfaction may not be enriching by conducting an offline
   survey with the customers about their satisfaction levels because it is biased by
   the most recent experience with the voice agent in the contact center and does
   not give information about the nature or level of dissatisfaction.

   A more accurate method to measure the customer dissatisfaction can be achieved
   by speech analytics by analyzing all the calls. Analysis of conversations can assist
   categorize 100 % of the calls by their root cause. These may include product fail-
   ure, a missed call-back commitment, incorrect billing, poor product, poor service
   quality, lack of agent knowledge, or wrongful service termination. Understanding
   the actual reasons ca help the enterprise identify exact improvements to enable
   customer satisfaction.

6. Customer Retention

   Enterprises that provide subscription-based services profit and benefit by customer retention because (a) acquiring a new customer is expensive, and payback
   is typically earned over a period that exceeds the initial term of service and (b)
   customers today have a wider choice of enterprises to challenging their current
   provider to match or beat a competitive offer.

   Speech analytics can enable spotting missed opportunities, gauge customer reac-
   tion to special offers, and compare the results of various retention strategies, so
   that the enterprise can take an informed judgement on the required agent training
   for higher customer retention.

7. Increasing Market Intelligence

   Market research with customers leads to acquisition of market intelligence, which
   is generally used by enterprises to make business decisions on products and other
   processes. Millions of pounds are spent annually to derive insights into the market. Traditionally, data suitable for market research is gathered from customers
   recollection of the interaction at a later date. Subsequently, the information provided by the customer is highly biased by later thoughts of the customer and very
   less dependent on the context. Also, there is a limit on the number of customers
   to be interviewed and how to choose them.

Speech analytics on the other hand, compared to traditional market research, is quantitative and verifiable because speech analytics can be applied across 100 gather data (instead of sampling a small percentage of calls), also it is timelier, with reports that can be generated in near real-time.

8. Monitoring Compliance

   Call centers are coming under increasing scrutiny and government regulations to maintain adequate standards of behavior of the agents and practice followed by them. While the call centers must necessarily comply with privacy restrictions, especially those that work under the auspices of the Financial Conduct Authority [4].

   Analyzing speech conversations provides a way to ensure that agents perform according to these preset operational expectations. Monitoring all the calls means there is potentially zero liability in terms of compliance on one hand and, on the other hand, this provides a feedback mechanism to help train and improve agent performance.

Clearly, there is a lot at stake for an enterprise if the large number of voice transaction between the agent and the customer happening every day were either not analyzed at all or analyzed partially. Observe that not all the derivable analytics requires understanding what is being spoken by the customer and the agent. For example, the logs maintained by the call center might be sufficient to give information about the average handling time per customer. However, a deeper analysis of the conversation between the agent and the customer will certainly give good and usable information.

References

1. L.S. Vygotskii, R.W. Rieber, M.J. Hall, in *The History of the Development of Higher Mental Functions*, The Collected Works of L.S. Vygotsky, ed. by Rieber, W. Robert, Cognition and Language: a Series in Psycholinguistics Series, (Plenum Press, New York, 1997)
2. F. Bell-Berti, K.S. Harris, Anticpatory coarticulation: Some implications from a study of lip rounding. J. Acoust. Soc. Am. **65**(5), 1268–1270 (1978). Status Report on Speech Research
3. M. Gavalda, J. Schlueter, *The Truth is Out There: Using Advanced Speech Analytics to Learn Why Customers Call Help-line Desks and How Effectively They Are Being Served by the Call Center Agent* (Springer, USA, 2010), pp. 221–243
4. Financial Conduct Authority. http://www.fca.gov.uk/

Chapter 2
Voice Analytics Process

2.1 Introduction

Analysis of a large volume of audio conversation is required to derive reliable analytics that is statistically significant. However, analyzing audio conversations manually is not only time-consuming and boring but also error prone. Automatic processing of audio conversations is the only option; this automatic mechanism of processing audio conversations to derive usable information is called Speech or Voice Analytics. So speech analytics can, at a very broad level, be defined as

> *The process of automatically deriving usable information by processing and analyzing a large quantum of statistically significant audio conversations between the customer and the agent to extract usable and actionable information.*

Deriving usable information from speech conversation is most often a two-step process (as seen in Fig. 2.1), namely that of

- Converting the audio conversation between the customer and the agent into text transcripts, followed by
- Analysis of the text transcripts, using natural language text processing techniques, to derive usable analytics.

While several other mechanisms exist, the most common one being that of spotting *key words* and *key phrases* in the audio conversation without completely transcribing the entire audio conversation into text [1]. The common process adopted by most commercial voice analytics tools available in the market [2–5] is that of converting the audio into text and then processing text to derive useful information, this is shown in Fig. 2.1.

A comprehensive picture of call center voice analytics process is sketched in Fig. 2.2. The process flow can be summarized as

Step 1: A user initiates the call which lands on the private branch exchange (PBX) of the call center. Note that the customers can call from any device (landline, mobile, VoIP), and from any environment (office, home, street). This

© The Author(s) 2015
S.K. Kopparapu, *Non-Linguistic Analysis of Call Center Conversations*,
SpringerBriefs in Electrical and Computer Engineering,
DOI 10.1007/978-3-319-00897-4_2

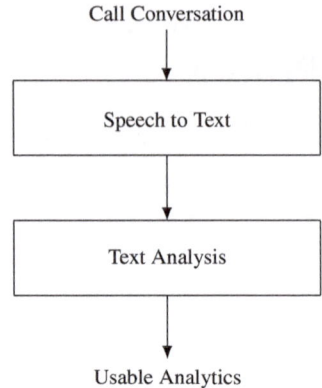

Fig. 2.1 Two-step voice analytics process

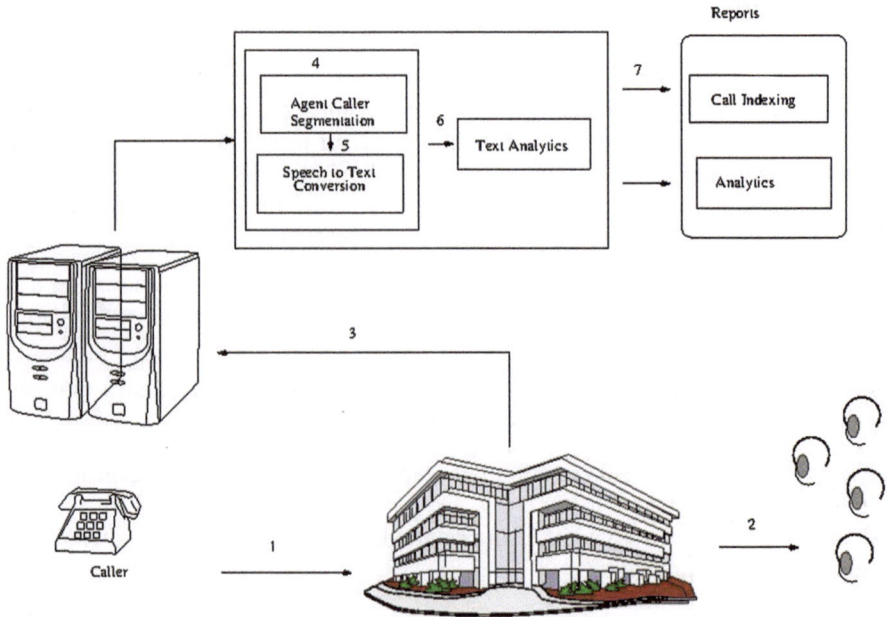

Fig. 2.2 Typical voice analytics process

is particularly important because the environment and the channel play a
prominent role in analyzing audio conversation.

Step 2: After waiting in the queue (more often than not), the call lands on the
agent's desk. Very often the wait time aggravates the customer satisfaction index [6] leading to an irate customer even before the customer starts
conversing with the agent.

Step 3: While the conversation is on, the call between the agent and the customer is being recorded along with some meta data (the number from where the call was made, the time of the day, etc.) on a server. These audio calls need to be analyzed for information to infer details being conversed. Typically, these calls are interlaced with speech of the customer, the agent and once in a while *hold music*.

Step 4.1: The interlaced speech is first processed to separate the hold music and voice.

Step 4.2: The voice portion is then separated into speech spoken by the agent and that spoken by the customer.

Step 5: Each of this speech segment is then passed through a speech to text converter, essentially an automatic speech recognition engine.

Step 6: The converted text is processed using text analysis tools to derive actionable and usable information.

Step 7: Capture the information embedded in the conversation which can be indexed for later search or generate reports.

So before the actual conversion of audio to text happens, there are two essential steps of preprocessing audio conversation, namely that of separating the hold music from voice and that of distinguishing the customer and the agent spoken speech. We look at this next.

2.2 Music Voice Separation

One of the important steps in analyzing call center conversation is the separation of voice and music. There are several approaches (see references in [7]) available to separate an audio which is interlaced by music and voice. One approach is to use nonlinear speech features like Teager energy [8] to obtain modulation-based features from an audio stream. These features can be used in a supervised learning scheme for segment-wise discrimination of vocal and music component in an audio stream.

Audio signal $x(t)$ is nonlinear, time-varying, and can be looked upon as a Amplitude Modulation–Frequency Modulation (AM–FM) model [8], namely

$$x(t) = a(t) \cos (\phi(t)) \tag{2.1}$$

where, $a(t)$ is the time-varying amplitude and $\phi(t)$ is defined as

$$\phi(t) = \omega_c t + \omega_m \int_0^t q(\tau)d\tau + \theta \tag{2.2}$$

where ω_c is the center frequency and ω_m is the maximum frequency deviation from the ω_c, $|q(t)| \leq 1$ and $\theta = \phi(0)$ is some arbitrary constant phase offset. The time-varying instantaneous angular frequency ω_i is defined as

$$\omega_i(t) \stackrel{def}{=} \frac{d}{dt}\phi(t) \tag{2.3}$$
$$= \omega_c + \omega_m q(t)$$

Equation (2.1) has both an AM and FM structure hence x(t) can be called an AM–FM signal.

It has been shown that this nonlinear modeling of speech helps in extraction of robust speech features [9]. Two different information signals can be simultaneously transmitted in the amplitude $a(t)$ and the frequency $\omega_i(t)$. The AM–FM model can be used to represent any speech signal $s(t)$ as a sum of AM–FM signals, namely

$$s(t) = \sum_{k=1}^{K} a_k(t)\cos(\phi_k(t)) \tag{2.4}$$

where K is the number of speech formants. Clearly, $a(t)$ and $\omega_i(t)$ for $k = 1, 2, \ldots K$ represents the speech signal $s(t)$. However, these have to be estimated from the speech signal.

Given a speech signal over some time interval, the problem is to estimate the amplitude envelope $|a(t)|$ and the instantaneous frequency $\omega_i(t)$ for each k and at each time t. One of the ways to estimate $a(t)$ and $\omega_i(t)$, is to first isolate individual resonance by bandpass filtering the speech signal around its formants and then estimating amplitude and frequency modulating signals of each resonance based on an *energy-tracking operator* as described in [10]. The Teager energy operator ψ (TEO) is defined as

$$\psi_c[x(t)] \stackrel{def}{=} \left[\frac{d}{dt}x(t)\right]^2 - x(t)\left[\frac{d^2}{dt^2}x(t)\right] \tag{2.5}$$

When ψ given by (2.5) is applied to the bandpass filtered speech signal (2.1), we get the instantaneous source energy, namely

$$\psi[x(t)] \approx a^2(t)\omega_i^2(t) \tag{2.6}$$

In the discrete form as is applicable to most speech processing systems [11], Eq. (2.5) can be written as

$$\psi[x[n]] = x^2[n] - x[n+1]x[n-1] \tag{2.7}$$

where $x[n]$ is the sampled speech signal.

The TEO is typically applied to a bandpass filtered speech signal, since its intent is to reflect the energy of the nonlinear flow within the vocal tract for a single resonant frequency. Although the output of the bandpass filter still contains more than one frequency component, it can be considered as an AM–FM signal, $r(t) = a(t)\cos(2\pi f(t)t)$. The TEO output of $r(t)$ can be approximated as

$$\psi[r(t)] \approx [a(t)2\pi f(t)]^2 \tag{2.8}$$

The AM–FM demodulation can be achieved by separating the instantaneous energy given in (2.6) into its amplitude and frequency components. ψ is the main ingredient of the first Energy Separation Algorithm (ESA) developed in [8] and used for signal and speech AM–FM demodulation, namely

$$f[n] \approx \cos^{-1}\left(1 - \frac{\psi[y[n]] + \psi[y[n+1]]}{4\psi[x[n]]}\right) \tag{2.9}$$

and

$$|a[n]| \approx \sqrt{\frac{\psi[x[n]]}{\left[1 - \left(1 - \frac{\psi[y[n]] + \psi[y[n+1]]}{4\psi[x[n]]}\right)^2\right]}} \tag{2.10}$$

where $y[n] = x[n] - x[n-1]$ and $f[n]$ is the FM component at sample n and $a[n]$ is the AM component at sample n. In practice, the speech signal is bandpass filtered using Gabor filters because of their optimal time-frequency discriminability [8], namely

$$s(t) = x(t) * g(t) \tag{2.11}$$

where $g(t)$ is given by

$$g(t) = \frac{1}{\sqrt{2\pi}\sigma}\left(e^{-\frac{t^2}{2\sigma^2}}\right)\left(e^{j(2\pi\omega_0 t)}\right) \tag{2.12}$$

where ω_0 is the center frequency and σ is the bandwidth of the Gabor filter.

In the case of audio signals, a Gabor filter-bank (placed at various critical band frequencies such as formant frequencies or at frequencies determined by Mel-scale)with a narrow bandwidth are used. The extraction of AM–FM components (2.9) and (2.10) from the bandpass filtered signal may be carried out using the Teager energy of the filtered signal. The efficiency of nonlinear speech features, namely instantaneous modulation features such as instantaneous amplitude and instantaneous frequencies have been studied for various applications like phoneme classification, speech recognition [9, 12], assessment of vocal fold pathology [13], stress detection [14], and music voice separation [7].

Figure 2.3 shows the typical distributions for vocal and music segments [7]. The instantaneous frequencies using three different Gabor filters ($\omega_0 = 240, 738, 1361$)

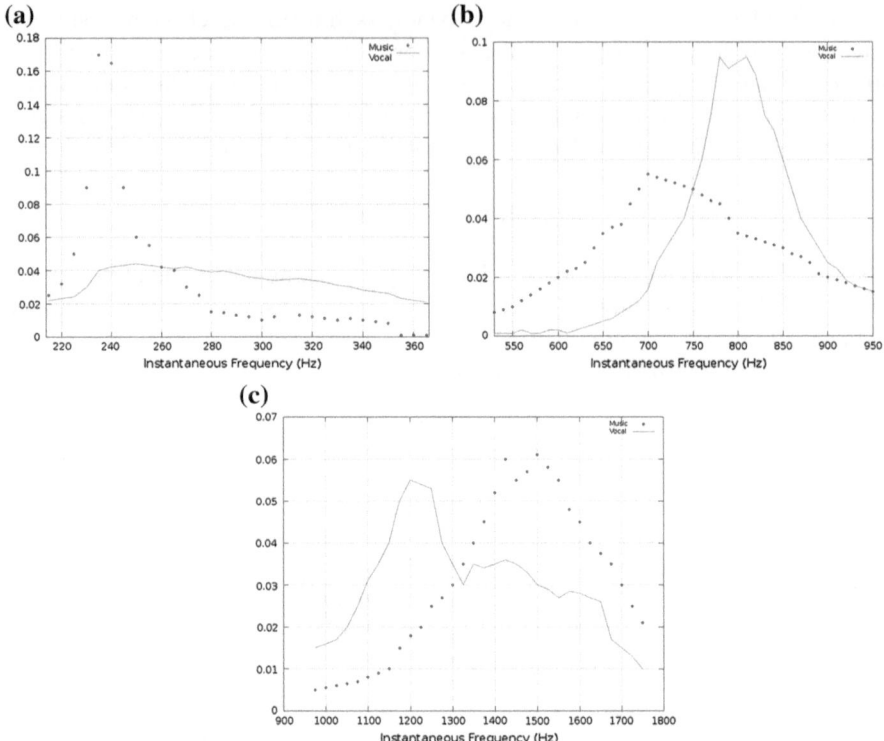

Fig. 2.3 Comparison of distribution of instantaneous frequencies in three bands for voice and music. **a** Band 1 (center frequency $\omega_0 = 240$). **b** Band 2 (center frequency $\omega_0 = 738$). **c** Band 3 (center frequency $\omega_0 = 1361$)

are extracted from the audio signal. The instantaneous feature distribution for voice and music segments of the audio, for three different bands, namely band 1 (center frequency $\omega_0 = 240$), band 2 (center frequency $\omega_0 = 738$) and band 3 (center frequency $\omega_0 = 1361$) are shown in Fig. 2.3a, b and c respectively.

It can also be seen from the distribution plots (Fig. 2.3) that the instantaneous frequency has very distinct distributions for voice and music segments in all the three frequency bands. This observation is exploited to distinguish voice and music components very reliably.

Once the music and voice in the audio conversation have been separated, the next task is that of distinguishing the voice spoken by the agent and that spoken by the customer. This is generally called speaker segmentation.

2.3 Agent Customer Speech Segmentation

A typical audio conversation available for processing and hence analysis is a conversation in *natural language* involving both the voice agent and the customer. For the purpose of meaningful analytics, it is mandatory that we know *who spoke what*, meaning we be able to distinguish the part of the audio conversation where the agent spoke and that part of the conversation where the customer spoke.

There are several audio segmentation techniques proposed in the literature for audio broadcast and group meetings [15–17], however, there is not much work done for telephone conversations [18] because telephone conversational speech has a much faster *speaker change rate*, large variation in speaking style of the speaker, presence of nonspeech sounds, crosstalk, and babble. Speaker segmentation is generally a multistep process consisting of

1. identifying the speaker change points in the audio conversation, followed by
2. identifying the number of speakers and then
3. associating each speech segment to a particular user.

Change point detection is, in general, the process of identifying change in some characteristics of the speech [19]. In a call center telephone conversation, the change in characteristics can be assumed to be that corresponding to the change from one person speaking to another. Typically, in telephone conversations, in addition to the noise and cross talk which is all too common, there are several instances when more than one person is speaking at the same time (talk over). Additionally, the conversation is skewed in the sense that the length or duration of one person speaking could be very small while that of the other person could be very large (a complaining customer). These are some of the challenges that one faces in telephone conversation compared to news audio broadcast where it is more controlled in terms of duration between changes in speakers.

A common process adopted for automatic change point detection in a telephone conversation is the use of a baseline change point detection technique [20]. The idea is to divide the entire speech into small overlapping windows of size 2–5 s; preprocessed each window to extract speech features like MFCC (Mel Frequency Cepstral Coefficient) and LSF (Line Spectral Frequency) [21] which are commonly used speech features for speaker segmentation [20] and then compare adjacent windows to identify change points in audio.

Though MFCC is extensively used in speech recognition; LSF speech parameters perform much better for speaker segmentation [22–24].

The comparison helps decide if the two adjacent windows of speech have similar characteristics or different, meaning if they originate from the same source or different sources. As seen in Fig. 2.4 the adjacent overlapping windows are represented by a red box and a blue box and are compared to determine if these adjacent windows of speech have similar characteristics or different. This process is applied on the entire speech signal by sliding along the time axis.

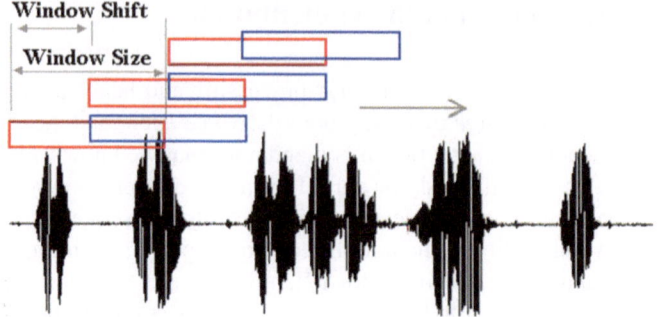

Fig. 2.4 Adjacent window comparison using overlapping windows

A speaker change point detection approach is described in detail for telephone conversation in [19]. It consists of a preprocessing step followed by identification of potential speaker change points followed by the selection of actual change points.

2.3.1 Two Pass Speaker Change Detection

The baseline change point detection technique used in speaker segmentation systems [25] is to divide the entire speech into small overlapping windows of size 25 s, then compare adjacent windows and decide whether the two adjacent windows of speech have similar characteristics or different, meaning if they originate from the same source or different sources. As shown using Fig. 2.4, the adjacent overlapping windows (represented by a red box and a blue box) are compared to determine if these windows of speech have similar characteristics of different. This decision is done by sliding along the time axis.

For a pair of adjacent windows being compared and found to be originating from different sources, the endpoint of the first window would be considered the change point. Clearly, the window size constrains the detection of short duration changes. Also, the localization (closeness of a detected change point to the actual change point) of the detected changes directly depends on the size of window overlap and the window shift [26].

2.3.1.1 First Pass: Coarse Detection

One can safely assume that the audio conversation, especially the call center conversation, starts with one speaker speaking for at least a few seconds (say *Speaker* 1, welcome message by the agent). These data are used to build a reference model λ_1 for *Speaker* 1. The rest of the audio conversation is analyzed using a sliding window (W) of length 2 s.

A hypothetical segment boundary is assumed at the center of the window with the first part (say W_1) of the window being assumed as a continuation of a speaker and the second part (say W_2) of the window being assumed as generated from the other speaker. A speech feature is extracted for the two sub-windows W_1 and W_2 under consideration using speech frames of length 20 ms with a frame overlap of 10 ms. Let there be N frames in each of the two sub-windows W_1 and W_2. We have,

$$W_1 = \{w_{11}, w_{12}, \ldots, w_{1N}\}$$

and

$$W_2 = \{w_{21}, w_{22}, \ldots, w_{2N}\}$$

The task is to decide if these two sub-windows belong to the same or different acoustic conditions and hence speakers. Assume that the hypothesis $I\!H_0$ indicates that the two sub-windows belong to one single multivariate Gaussian process or a single speaker, namely

$$I\!H_0 : W_1, W_2 \sim N(\mu_W, \Sigma_W) = N_W$$

The hypothesis $I\!H_1$ indicates that the two segments (W_1, W_2) are generated by two different multivariate Gaussian processes or two speakers, namely,

$$I\!H_1 : W_1 \sim N(\mu_{W_1}, \Sigma_{W_1}) = N_{W_1} \quad \text{and}$$

$$W_2 \sim N(\mu_{W_2}, \Sigma_{W_2}) = N_{W_2}$$

where μ_W, μ_{W_1} and μ_{W_2} are the mean vectors and Σ_W, Σ_{W_1}, and Σ_{W_2} are the covariance matrices of the entire window W and the two sub-windows W_1, W_2 respectively. The generalized log likelihood ratio (GLR, $I\!R_W$) between the hypotheses $I\!H_0$ and $I\!H_1$ for the window W is defined as

$$I\!R_W = \log L(W, N_W) - (\log L(W_1, N_{W_1}) + log L(W_2, N_{W_2})) \qquad (2.13)$$

The GLR ($I\!R_W$) is computed [27] for a pair of adjacent sub-windows of same size and the analysis window is then shifted by a step length of 0.5 s along the speech signal and the likelihood ratio for the new window is computed. Negative $I\!R_W$ indicates that the sub-windows are better represented by N_{W_1} and N_{W_2} rather that the whole window W being represented by N_W meaning W had a speaker change point. The GLR distances thus computed for all windows for the audio track are computed and a threshold T (chosen empirically through experiments) is used to detect all possible candidate speaker change points. The threshold is so chosen, so that one does not miss out on any actual change points.

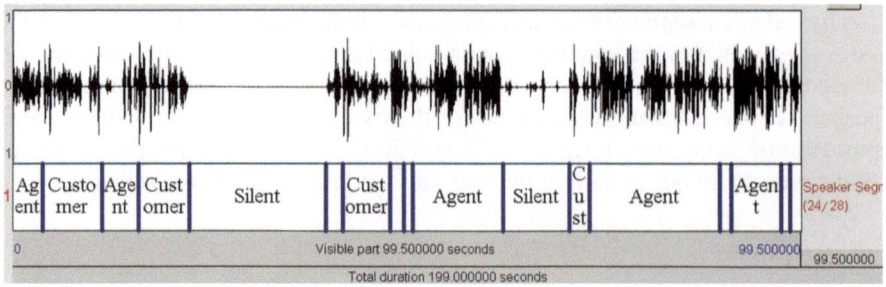

Fig. 2.5 Audio conversation with speaker segmentation

2.3.1.2 Second Pass: Speaker Change Detection

A second-level analysis is carried out to narrow down on the most likely speaker change point identified in the first pass. The candidate speaker change points detected in the above step are considered sequentially and we try to find the pair of segments that have a high likelihood of being from different speakers. For this, we use the initial part of the data which is assumed to be from *Speaker* 1 having a model parameter set represented as, λ_1. The similarity of the sub-window $W_1 = [w_1, w_2, \ldots, w_N]$ in the neighborhood of a candidate speaker change point detected in the first pass is computed as the log-likelihood

$$S(W_1|\lambda_1) = \sum_{i=1}^{N} \log(p(w_i|\lambda_1)) \tag{2.14}$$

The acoustic probability that an observed feature vector w_i was generated by the model, λ_1 is given by,

$$p(w_i|\lambda_1) = \frac{1}{\sqrt{(2\pi)^d|\Sigma|}} exp\left\{-\frac{1}{2}\left((w_i-\mu)^T \Sigma^{-1}(w_i-\mu)\right)\right\} \tag{2.15}$$

where d is the dimension of the feature vector w_i. Since true densities of the *Speaker* 1 is unknown, they can be approximated by sample mean and variances for computing the speaker model λ_1. The change points for which the adjacent sub-window segments have a large difference in similarity computed from (2.14) are identified as valid speaker change points. The first valid speaker change point signals the beginning of a second speaker segment. The first speaker model parameters are modified using all the frames up to the detected change point.

Figure 2.5 shows a sample speech conversation after speaker segmentation. Once this is done, a speech to text conversion (Fig. 2.6a) would result in a corresponding text transcription as shown in Fig. 2.6b. However, there are challenges in the process of converting speech to text.

(a)

| Welcome to XYZ bank- |
| ing services how may I |
| help you I need infor- |
| mation about my credit |
| card spend please hold |
| on sure can you confirm |
| your last name and ad- |
| dress please it is matt |
| mukund and the address |
| is 208 gir apartments la. |

(b)

| S1: Welcome to XYZ banking services how may I help you |
| S2: I need information about my credit card spend please |
| S1: hold on |
| S2: sure |
| S1: can you confirm your last name and address please |
| S2: it is matt mukund and the address is 208 gir apartments la. |

Fig. 2.6 Sample transcription. **a** Without and **b** with speaker segmentation

2.4 Speech to Text Conversion

The process of speech to text conversion is commonly know as automatic speech recognition (ASR). It should noted that like most AI systems, the process of building a speech recognition system is based on learning, namely the recognition system needs to be taught what it is required to perform. There are essentially (see Fig. 2.7) three main blocks involved in this recognition process. The first one is the acoustic models (AM), the second is the lexicon and the third is the statistical language model (SLM).

The acoustic model (AM) is an important component of speech to text conversion process, it models various sounds (phonemes) that make up a spoken word. In general, AMs are statistical models and is trained using large amount of speech corpus.

Speech corpus is a carefully constructed data set which consist of the spoken acoustic data plus a manually transcribed text corresponding to the audio data.

The lexicon (or the pronunciation dictionary, see Fig. 2.8) contains a list of words that should be recognized by the ASR engine. There are different approaches to construct lexicons using grapheme to phoneme mapping techniques [28, 29], however,

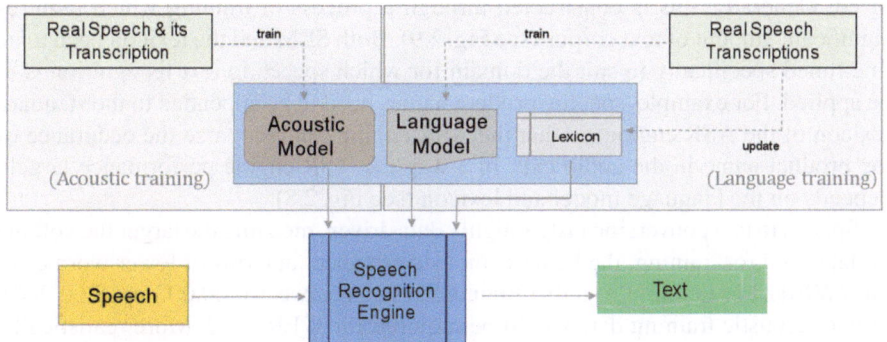

Fig. 2.7 Automatic speech recognition process

Fig. 2.8 A sample Lexicon
that assists ASR

Word	Phonemic Sequence
ALLIANZ	AE L IY AH N Z
INSURANCE	IH N SH UH R AH N S
POLICY	P AA L AH S IY
SURRENDER	S ER EH N D ER
SECURITY	S IH K Y UH R AH T IY
SUPERVISOR	S UW P ER V AY Z ER
TAX	T AE K S
TRANSACTION	T R AE N Z AE K SH AH N

Fig. 2.9 A sample text corpus
used to build SLM

```
THANKS FOR CALLING THIS IS MIKE CAN
WE START WITH A POLICY NUMBER POLICY
NUMBER OOH LET ME FIND THAT REAL QUICK
OK POLICY NUMBER EIGHT FIVE ZERO FIVE
TWO ONE FOUR AND THE CUSTOMER IT'S
BEVERLY B ROCKWOOD CAN YOU TRANSFER
TO THE CUSTOMER PLEASE YA GRANNY YOU
GET ON THE PHONE HELLO HEY HOW YOU
DOING MAAM YOUR NAME PLEASE MY NAME
IS BEVERLY B ROCKWOOD THANK YOU BEVERLY
CAN YOU VERIFY THE ADDRESS WE HAVE N
FILE FOR YOU PLEASE FOURTEEN TWENTY
FOUR EAST VIRGINIA AVENUE AND THE LAST
IS THE POST OFFICE BOX NUMBER IT'S THE
POST OFFICE BOX NUMBER OH OK MY HUSBAND
DID THAT WAY HAT'S THE REASON WHY I AM
KIND OF LANKY AN YOUR ADDRESS WITH THE
POST OFFICE BOX AGAIN
```

A lexicon, in general is handcrafted by a Linguistic and takes into account the actual manner in which a particular word is Pronounced. It should be noted that the same word can have more than one pronunciation.

Statistical Language Model or just the language model (LM) models the syntax and grammar of the sentences that are expected to be spoken and hence to be recognized. Generally, this is constructed through a process of training which requires significant amount of text corpus (see Fig. 2.9). Both SLM and the lexicon need to be fine-tuned specifically to suit the domain for which speech to text recognition is to be applied. For example, specific product names need to be appended to the standard lexicon of the ASR engine, so that the ASR engine can recognize the occurance of the product name in the audio call. In a way the ASR engine performance largely depends on the language model and lexicon (see Fig. 2.8).

Speech to text conversion task is highly data driven, meaning the larger the volume of data used for training, the better is the performance. in terms of lower word error rate (WER). For example, a study [30] indicates that between 6,00,000 and 8,00,000 h of data acoustic training data would be required for WER $\rightarrow$ 0. More realistically, as discussed in [31] WER as low as 17 % could be achieved on voice search queries

Table 2.1 Types of error in ASR

Actual	A	B	C	D	*
Hypothesis	*	B	S	D	E

in English by using SLM trained on 230 billion words. However, it should be noted that apart from the amount of training data, the type of training data matters a lot in terms of the training data resembling the actual test scenario. Resemblance in terms of environmental conditions.

The performance of an Speech to text conversion process is generally measured as Word Error Rate (W E R),

$$WER = \frac{\#\ of\ word\ errors}{\#\ of\ reference\ words}$$

where # of word errors is the sum of number of word deletions, word insertions and word substitutions required to match the output of a Speech to text hypothesis to the actual speech utterance.

We see above (in Table 2.1) that there are three types of word errors. The word A is not recognized at all (Deletion error), C is recognized as S (Substitution error), E is falsely recognized when it is not actually present in the speech (Insertion error). For the above example, $WER = 3/4$. $WER \rightarrow 0$ indicates better accuracy WER.

The current adopted method in most commercial voice analytics tools (for example, [2–5]) for analyzing call center conversations, as mentioned earlier, is one of converting speech to text followed by natural language text processing. However, there are several challenges in using this two-step approach to analyze call center conversations.

2.4.1 Speech-to-Text Conversion Challenges

There are a significant number of challenges that crop up during the process of converting speech to text, especially for natural language spoken conversations on the telephone channel.

As observed earlier, building a speech recognition system to automatically transcribe naturally spoken call center conversation requires extensive training, not only in terms of construction of a SLM and a lexicon but also in terms of training acoustic models.

A language is said to be resource rich if it has a reasonable quality and quantity of speech corpus. Else the language is categorized as being Deficient.

Unfortunately, the state of the art, even for a resource rich language like English is poor because of use of slangs, different accents, different environmental conditions to name a few. Post extensive training, a well-trained speech recognition engine even for a resource rich language like English, the speech recognition performance on a natural language spoken conversation has a *WER* of about 40 − 50 % (see Appendix F).

(b)

(a)

SRI LANKA GETS USDA APPROVAL FOR WHEAT PRICE Food Department officials said the U.S Department of Agriculture approved the Continental Grain Co sale of 52,500 tonnes of soft wheat at 89 U.S Dlrs a tonne C and F from Pacific Northwest to Colombo They said the shipment was for April 8 to 20 delivery

... Srilankan gets you s.d. approval for wheat thrice fire department officials said the u.s. department of agriculture approve p continental crane co sell off fifty two thousand five funded tons office of to be at each team nine new west dollars a tan c and as from pacific northwest to colombo they said the shipment was for april eke to twenty deliver

Fig. 2.10 Example of noise due to speech to text process. **a** Actual text and **b** ASR transcribed text

Table 2.2 Factors impeding Speech to text

Category	Parameters
Source/Channel	Background noise, speech codec, microphone, telephone
Linguistic	language, dialect, mixed language, accent, pronunciation, domain, vocabulary size, proper-names, OOV
Non-linguistic	Spoken speech (anger, hesitation, fast)
Operational	real time, offline, multi-pass ASR

Subsequently, irrespective of the type of text analysis approach adopted, the analysis of the text is affected by the noise in the transcription. This affects the usable information derived from these noisy transcriptions. There is a relatively new area of work concentrating on noisy text analysis [33–35].

ASR performance even under controlled conditions, like in [32], where an original article was read by a single user, the speech-to-text conversion accuracy is poor. Figure 2.10a shows the original text, while 2.10b shows the output of an ASR. Clearly, the ability to automatically transcribe a natural language speech is very poor. As noted earlier, the speech-to-text conversion mechanism is based on the process of learning. If the training data, in this case speech, is for a particular channel or a particular environment, then the speech to text processing fails if there is a variation in these environmental conditions. Hence, the presence of environmental noise in the conversational speech is a further impediment in the speech to text conversion process [36] and so is the variation in the channel characteristics. A more complete list is captured in Table 2.2.

The speech-to-text conversion process for resource deficient languages is further compounded because of lack of availability of speech corpus for training (acoustic models, statistical language models, lexicon, see Fig. 2.7) and hence the recognition accuracies are very poor. However, speech corpus is a central element for training the acoustic models used in a speech recognition engine. Additionally, constructing a speech corpus for a language is an expensive, time-consuming, and laborious process [37]. Appendix E details on how to construct a speech corpus for a resource deficit language.

Building Speech recognition application for resource deficient languages is a challenge because of the unavailability of a speech corpus.

A mechanism to build an inexpensive speech corpus, for resource- deficient languages Indian English and Hindi, by exploiting existing collections of online speech data to build a Frugal speech corpus is proposed in [37].

For resource-deficient languages the two-step process (steps of speech to text followed text analysis) of analyzing call center conversation to derive analytics does not exist. However, there has been a segment of work that makes use of existing language resources to build speech recognition engine for a resource deficient language (see for example [37–41]). Additionally, the speech to text conversion process is further complicated in a multilingual country, like India, where people tend to use more than one language in the same sentence. This is called mixed language use [42] and is also known as code switching in literature. Speech-to-text conversion of such speech is highly poor because most often the speech recognition engines are designed with the assumption that users stick to one language during their conversation. This poses problems in automatic recognition of linguistic content of the conversation.

These challenges in linguistic processing of spoken speech make it difficult to use the two-step process to analyze call conversations, thus opening up options to use non-linguistic processing of speech to derive usable information. For resource deficient languages one needs to necessarily rely on techniques that do not require the linguistic speech to text conversion process for such languages.

Additionally, reasons to adopt non-linguistic processing could stem from the fact that sometimes it is just sufficient to know in a call center setting which call was abnormal; without the need to know what was the explicit (linguistic) reason for abnormality.

References

1. M. Gavalda, J. Schlueter, The truth is out there: Using advanced speech analytics to learn why customers call help-line desks and how effectively they are being served by the call center agent, in *Advances in Speech Recognition*, ed. by A. Neustein (Springer, USA, 2010), pp. 221–243
2. Nexidia. http://www.nexidia.com/solutions/contact_center/product_suite
3. Utopy. http://www.utopy.com/index.php?page=intelligent-script-adherence
4. Verint. http://verint.com/contact_center/section2a.cfm?article_level2_category_id=21& article_level2a_id=343
5. Nice. http://www.nice.com/smartcenter-suite/interaction-analytics
6. I. Ahmed, S.K. Kopparapu, Interactive voice response mashup system for service enhancement. Recent Patents on Telecommun. **1**, 100–108 (2012)

7. S.K. Kopparapu, M.A. Pandharipande, and G. Sita. Music and vocal separation using multiband modulation based features. In: IEEE Symposium on Industrial Electronics Applications (ISIEA), pp. 733–737 (2010)
8. P. Maraso, J.F. Kaiser, T.F. Quatieri, Energy separation in signal modulations with applications to speech analysis. IEEE Trans. Signal Proc. **41**, 3024–3051 (1993)
9. D. Dimitriadis, P. Maragos, A. Potamianos, Robust am-fm features for speech recognition. IEEE Signal Process. Lett. **12**, 621–624 (2005)
10. T.F. Quatieri, T.E. Hanna, G.C. O-Leary, Am-fm separation using auditory-motivated filters. IEEE Trans. Speech and Audio Proc. **5**, 465–480 (1997)
11. H.L. John, H.G. Zhou, J.F. Kaiser, Nonlinear feature based classification of speech under stress. IEEE Trans. Signal Proc. **9**, 203 (2001)
12. D. Dimitriadis, P. Maragos, Continuous energy demodulation methods and application to speech analysis. Speech Communication **48**, 819–837 (2006)
13. L.G.C. John, J.H.L. Hansen, J.F. Kaiser, Vocal fold pathology assessment using am autocorrelation analysis of the teager energy operator. In in Fourth Int. Conf. Spoken. Language **2**, 757–760 (1996)
14. R. Mandar and H. John. Frequency distribution based weighted sub-band approach for classification of emotional/stressful content in speech. In:EUROSPEECH-2003, pp. 721–724 (2003)
15. G. Rashmi, N. Balakrishnan, A novel method for two-speaker segmentation. In: INTERSPEECH-2004, pp. 2337–2340 (2004)
16. A.N. Iyer, U.O. Ofoegbu, R.E. Yantorno, S. Wenndt. Speaker modeling in conversational speech with application to speaker count. In Proceedings of ICSLP (2006)
17. S. Soni, I. Ahmed, S.K. Kopparapu. Automatic segmentation of broadcast news audio using self similarity matrix. CoRR, abs/1403.6901 (2014)
18. S.K. Kopparapu, A. Imran, G. Sita. A two pass algorithm for speaker change detection. In TENCON 2010–2010 IEEE Region 10 Conference, pp. 755–758 (2010)
19. I. Ahmed ,S. Kopparapu. Speaker change detection in telephone speech. In International Conference on Signals, Systems and Automation, Vallabh Vidyanagar, India (2009)
20. S.E. Tranter, D.A. Reynolds, An overview of automatic speaker diarization systems. IEEE Trans. Audio, Speech Lang. Process. **14**, 1557–1565 (2006)
21. Y. Tianren, X. Juanjuan, Lu Wei. The computation of line spectral frequency using the second chebyshev polynomials. In 6th International Conference on Signal Processing, pp. 190–192, vol 1 (2002)
22. P. Delacourt, C.J. Wellekens, Distbic: A speaker-based segmentation for audio data indexing. Speech Communication **32**, 111–126 (2000)
23. Lie Lu ,H.-J. Zhang, Real-time unsupervised speaker change detection. In Proceedings to 16th International Conference on Pattern Recognition, pp. 358–361 (2002)
24. A.G. Adami, S.S. Kajarekar, H. Hermansky. A new speaker change detection method for two-speaker segmentation. In Proceedings of IEEE International Conference on Acoustics, Speech, and Signal Processing -ICASSP 02, pp. IV-3908 - IV-3911 (2002)
25. S.E. Tranter, D.A. Reynolds, An overview of automatic speaker diarization systems. IEEE Trans. Audio Speech Lang. Process. **14**, 1557–1565 (2006)
26. I. Ahmed, S.K. Kopparapu, Speaker change detection in telephone speech. In International Conference on Signals, Systems and Automation, Vallabh Vidyanagar, India (2009)
27. B. Narayanaswamy, G. Rashmi, R. Stern, *Voting for two speaker segmentation* (In Proc, ICSLP, 2006)
28. Maximilian Bisani, Hermann Ney, Joint-sequence models for grapheme-to-phoneme conversion. Speech Commun. **50**(5), 434–451 (2008)
29. M. Laxminarayana, S. Kopparapu, Semi-automatic generation of pronunciation dictionary for proper names: an optimization approach. Proceedings of the 6th International Conference on Natural Language Processing (ICON08), pp. 118–126 (2008)
30. K. M. Roger, A comparison of the data requirements of automatic speech recognition systems and human listeners. EUROSPEECH'03, pp. 2582–2584 (2003)

31. C. Ciprian, B. Dan, S. Maria, N. Patrick, K. Shankar, Large scale language modeling in automatic speech recognition. Technical report, Google Techreport (2012)
32. Reuters Transcribed Subset. http://kdd.ics.uci.edu/databases/reuters_transcribed/reuters_transcribed.html. Viewed July 08, 2013
33. J. Mamou, D. Carmel, R. Hoory, Spoken document retrieval from call-center conversations. In Proceedings of the 29th annual international ACM SIGIR conference on Research and development in information retrieval, SIGIR '06, pp. 51–58, New York (2006)
34. D. Lopresti, S. Roy, K. Schulz, L.V. Subramaniam, Special issue on noisy text analytics. International J. Document Anal. Recognit. (IJDAR) 14(2), 111–112 (2011)
35. S. Agarwal, S. Godbole, D. Punjani, S. Roy, How much noise is too much: A study in automatic text classification. In Seventh IEEE International Conference on Data Mining, ICDM 2007. pp. 3–12 (2007)
36. M. L. Narayana, S.K. Kopparapu, Effect of noise-in-speech on mfcc parameters. In Proceedings of the 9th WSEAS international conference on signal, speech and image processing, and 9th WSEAS international conference on Multimedia, internet and video technologies, SSIP '09/MIV'09, pp. 39–43, Stevens Point, Wisconsin, USA, 2009. World Scientific and Engineering Academy and Society (WSEAS)
37. I. Ahmed, S.K. Kopparapu, Speech recognition for resource deficient languages using frugal speech corpus. In IEEE International Conference on Signal Processing, Communication and Computing (ICSPCC), pp. 750–755 (2012)
38. O. Cetin, M. Plauche, U. Nallasamy, Unsupervised adaptive speech technology for limited resource languages: A case study for tamil. In: Proceedings of International Workshop on Spoken Languages Technologies for Under-resourced languages (SLTU), Hanoi, Vietnam (2008)
39. M. Plauche, O. Cetin, U. Nallasamy. How to build a spoken dialog system with limited or no language resources. In AI in ICT4D. ICFAI University Press, India (2008)
40. T. J. Arnar, W. D. Edward, I. Koji, S. Furui. Language model adaptation for resource deficient languages using translated data. In INTERSPEECH'05, pp. 1329–1332 (2005)
41. I. Dawa, Y. Sagisaka, S. Nakamura, Investigation of asr systems for resource-deficient languages. ACTA AUTOMATICA SINICA 1, 1–8 (2008)
42. K. Bhuvanagiri, S.K. Kopparapu, Mixed Language Speech Recognition without Explicit Identification of Language. American J. Signal Process. 2, 92–97 (2012)

Chapter 3
Call Center Linguistic Analytics

Voice-based call centers enable customers query for information by speaking to human agents. Most often these call conversations are recorded by call centers because (a) they are mandated by the law of the land and, more importantly, (b) with the express intent of trying to identify hidden information in these conversations that can help improve the performance of the call center to serve the customer better.

Call center recorded customer telephonic interactions are a rich source of information begging to be identified, extracted, and put to use. Information in terms of customer opinion and sentiment about a recently introduced product or a newly adopted strategy, services, process, and operational issues can be extracted. Additionally, these calls carry information about the voice agent performance which can be used for targeted and personalized training of the agent.

Speech Analytics (SA) or Call Center Analytics is the method of automatically analyzing recorded telephone calls to extract useful and usable information. An accurate analysis of the call center conversations shed light on some very crucial usable information that would otherwise be lost [1].

3.1 Spotting Problematic Calls

Analyzing conversations is an expensive task if done manually; in addition, the manual task is error prone and very often not comprehensive. Even for a simple task of identifying all the problematic conversations recorded between the agent and the customer would require 100 % of the conversations to be listened, from the beginning to the end of the call by a human supervisor to determine if the conversation was normal or problematic. This listening from end to end without skipping any part of the audio conversation completely is mandatory if one is keen on identifying *all* the problematic calls without missing a single problematic call.

© The Author(s) 2015
S.K. Kopparapu, *Non-Linguistic Analysis of Call Center Conversations*,
SpringerBriefs in Electrical and Computer Engineering,
DOI 10.1007/978-3-319-00897-4_3

Fig. 3.1 A normal
conversation

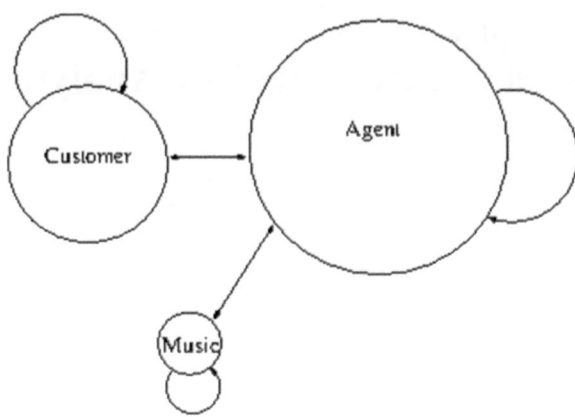

*The structure or the anatomy of a normal call is characterized by interlaced spoken seg-
ments between the agent and the customer with fewer instances of (hold) music. An ideal
conversation starts with the agent and ends with the agent and has the agent speak more
than the customer. Figure 3.1 shows a graphical representation of a normal call. The nodes
represent the speaker (agent, customer, hold music) and the size of the node represents the
spoken content in terms of the amount of time spoken.*

If there are N calls, each of average duration t seconds, then a total of $N \times t = T$
hours of call recordings exist. One needs to listen to the T hours of speech completely
to identify all the problematic calls (represented by the large circle in Fig. 3.2).

However, the *actual* portion which signifies abnormality within an audio conversation is of small duration and is represented by the small dark circles within the
problematic calls. Typically, driven by practical reasons, only a very small fraction
of all the recorded call conversations, selected in random, are carefully heard by a
human supervisor (marked by the hashed region).

It can be observed that a bad choice of conversations (say any call outside the large
circle in Fig. 3.2) selected to be heard by the supervisor will lead to the supervisor
missing out on identifying problematic calls altogether. In Fig. 3.2 we show for
purposes of representation 16 different selection of the T hours of conversation.
Clearly, 12 out of 16 selections would lead to missing out on listening to any portion
of the conversation that is likely to be abnormal or problematic. Clearly, there is
a large portion of the recorded calls that is not part of the calls listened to by the
supervisors and hence more often than not the problematic calls might not be part of
the human analysis.

*Landing on spotting a problematic call manually is like searching for the small black dots
in a maze (Fig. 3.2).*

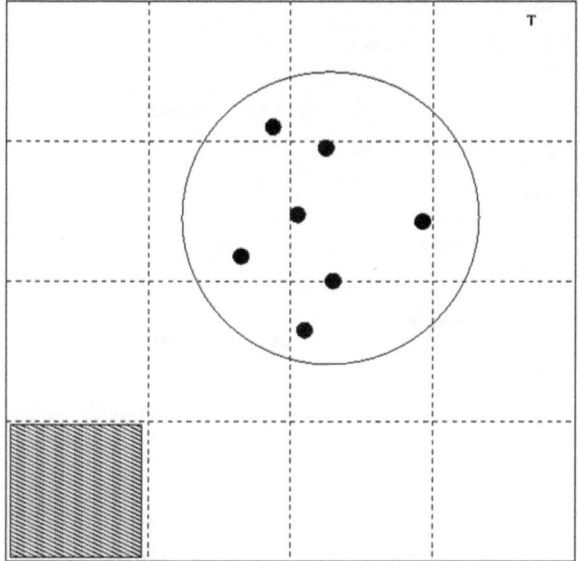

Fig. 3.2 Call conversation with problems; spotting the *dark dots* in a huge maze

3.1.1 Detailed Analysis

The practical difficulty in identifying problematic calls manually is depicted in
Fig. 3.3 [2]. The call duration is shown on the x-axis while different calls are depicted
along the y-axis. As seen in Fig. 3.3, the calls are of different duration and the gray
color depicts the actual length of each call. The black color within a call displays
the location of a problem situation during the call. Clearly, the location of problem
situation is arbitrary and typically the duration of the problem situation is also very
small.

Assume that there are N calls that the supervisor has to analyze and flag as being
either normal or problematic. Let d_i denote the duration (in seconds) of the ith call.
For a supervisor to actually identify all the calls with a problem, he would have to
listen to $\sum_{i=1}^{N} d_i$ seconds of conversation which in general is impractical in most
situations. Today, supervisors randomly select a small set of calls, say some $k\%$ of
N, namely

$$M = \frac{kN}{100},$$

where $M \ll N$. For each of these m calls ($m = 1, 2, \ldots M$), the supervisors would
randomly select a point in time to listen to the call, say l_{start}^m and listen till l_{stop}^m, namely
for a duration of

$$L_m = l_{\text{stop}}^m - l_{\text{start}}^m.$$

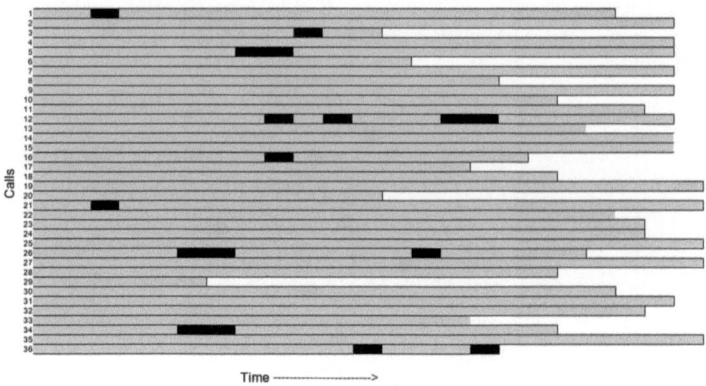

Fig. 3.3 Call conversation with problems. The *dark portions* represent regions of problem in the call

This reduces his listening time to $\sum_{m=1}^{M} L_m$ which is more practical and can be handled by a human supervisor.

Observe that $\sum_{m=1}^{M} L_m \ll \sum_{i=1}^{N} d_i$ because $M \ll N$ and $L_m \ll d_i$.

Suppose there are μ out of N calls that are problematic, the probability (P_{all}) that the supervisor exactly chooses those μ calls for listening is

$$P_{all} = \frac{1}{{}^{N}C_{\mu}} = \frac{(N-\mu)!\mu!}{N!}$$

$$= \frac{\mu!}{\Pi_{l=0}^{\mu-1}(N-l)} \qquad (3.1)$$

Figure 3.4 shows the log probability $(log(P_{all}))$ of identifying all the problematic calls correctly, from the set of N calls, as a function of the number of calls (μ) being actually problematic.

Clearly, if the number of problematic and normal calls are equally probable then the probability of identifying all the problematic calls exactly is low (represented by the trough in Fig. 3.4). Generally, the number of problematic calls is about 5–15% of the total number of calls, while the probability is better than when the problematic calls are close to 50% of the total number of calls, it is still very small.

If $N = 100$ and $\mu = 10$ then we get P_{all} the probability of spotting all the problematic calls is 5.7769×10^{-14}, which is very small.

For this reason, the probability of a supervisor being able to *pick up* all the problematic calls is very small; as a result most problematic calls go unnoticed for any corrective action to be taken. There is a need for automatic analysis of call conversations to spot the problematic calls.

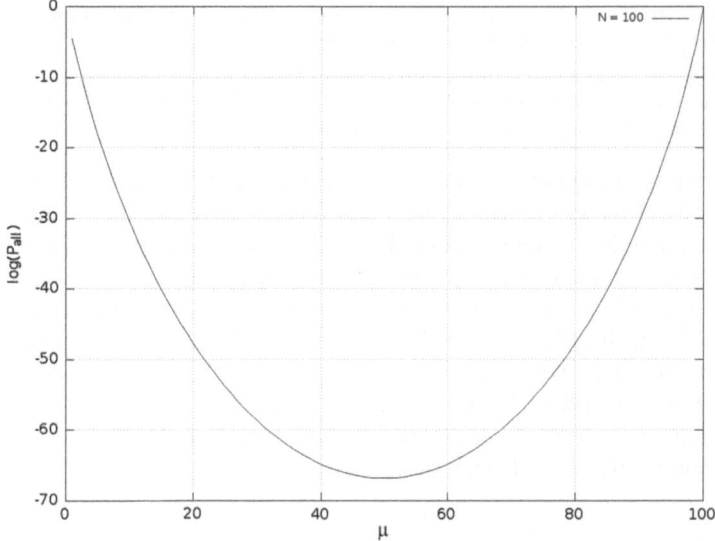

Fig. 3.4 Log probability of identifying all the problematic calls

3.2 Automatic Spotting

Using automatic speech recognition (ASR) technology, the task of identifying problematic calls reduces to one of automatic transcription of call conversation followed
by analyzing the words or phrases in the text transcriptions [3, 4]. As stated earlier,
the process of converting audio conversations into transcribed text is far from accurate, even for a resource rich language, even if one uses the state of the art speech
recognition technology. The recognition accuracies further deteriorate when there is
no readily available ASR for the language spoken in the conversation [5].

In many cases, even an accurate transcription of the audio conversation is not
sufficient to identify a problem situation in a call conversation [6] because the problem
in the call might not translate into meaningful transcribed text. For example, /thank
you/ spoken in a sarcastic tone might not suggest a problem in the call because
the phrase *"Thank You"* is generally associated with a satisfied customer and hence
would be categorized as a non-problematic call.

As mentioned earlier, the usual way of deriving usable information from speech
is through the process of speech to text followed by text to analytics. However, this
two-step analysis has the following limitations:

- *Need for a language-specific speech recognition engine.*
- *Need for a well-trained and robust speech recognition engine.*
- *Need for a rich speech corpus, which often does not exist.*
- *Speech recognition accuracies even for resource rich languages is of the order of* 50–60 %
  for natural conversations.

- *Even a 100% speech-to-text conversion would not be able to distinguish words that suggest positive sentiment but spoken in a negative way.*
- *Most speech-to-text processing engines assume the use of single language; but natural conversations are far from this especially in a multilingual country like India where bilinguality due to modernization is prominent.*

Also, many a time the need in the call center is more in terms of quality monitoring [7, 8] of the conversation [9] which does not need lexical interpretation of the conversation. For example, Ref. [9] describes a mechanism to observe the quality of calls in a call center environment. The system allows monitoring of several live ongoing audio conversations to alert supervisor regarding vulnerable deviations in the call being handled by a call center agent. The alerts are displayed on the supervisor console by way of providing graphical visual display in order to seek the attention of the supervisor. In this case and in several other situations, there is a need for non-linguistic processing of the speech conversations to keep it grossly language independent enabled speech analytics.

References

1. C. Bailor, The why factor in speech analytics. CRM Magazine, August 2006
2. M.A. Pandharipande, S.K. Kopparapu, A language independent approach to identify problematic conversations in call centers. ECTI Trans. Comput. Inf. Technol. **7**(2), 146–155 (2013)
3. G. Mishne, D. Carmel, R. Hoory, A. Roytman, A. Soffer, Automatic analysis of call-center conversations. *in Proceedings of the 14th ACM international conference on information and knowledge management*, CIKM '05, (2005), pp. 453–459, New York
4. VERINT. Speech analytics essentials for audiolog, http://www.verint.com/solutions/enterprise-workforce-optimization/products/speech-analytics/impact-360-speech-analytics-essentials-for-audiolog
5. S.K. Kopparapu, I. Ahmed, Enabling rapid prototyping of an existing speech solution into another language. *in 14th Oriental COCOSDA Conference*, October 2011
6. F. Cailliau, A. Cavet, Mining automatic speech transcripts for the retrieval of problematic calls. *in Proceedings of the 14th international conference on computational linguistics and intelligent text processing - Vol. 2*, CICLing'13, (Springer, Berlin, 2013), pp. 83–95
7. G. Zweig, O. Siohan, G. Saon, B. Ramabhadran, D. Povey, L. Mangu, B. Kingsbury, Automated quality monitoring for call centers using speech and nlp technologies. *in Proceedings of the 2006 conference of the north american chapter of the association for computational linguistics on human language technology: companion volume: demonstrations*, NAACL-demonstrations '06, (2006), pp. 292–295, Stroudsburg
8. A. Pande, S.K. Kopparapu, System for conversation quality monitoring of call center conversation and a method thereof. US Patent App. 13/742,829, 15 August 2013
9. A. Pande, S.K. Kopparapu, A system for conversation quality monitoring of call center conversation and a method thereof. Indian Patent 343/MUM/2012 (2012)

Chapter 4
Non-linguistic Speech Processing

There has been an increase in spoken interaction between people from different geographies or different cultural backgrounds prominently in the call center scenario. Noticeably, ineffectiveness of conversations is prominent when two people from different cultures converse in a language common to them. One of the main reasons for conversation ineffectiveness is driven by the *way* the language is spoken and not so much by *what* is being spoken.

The idea of non-linguistic speech processing also stems from the fact that there is a need for language-specific resources to process a speech signal to extract the spoken linguistic content in it. In addition, there is an inherent assumption that the language being conversed is known a priori for the purpose of converting speech into text. Extensive training is required to enable good speech-to-text conversion accuracies; training requires loads of training speech corpus. In the previous chapters, we have seen that because of several factors the availability of such a corpus is small for several resource deficient languages. Additionally, with bilinguality due to modernization, even the availability of a speech processing platform to convert speech into text is questionable because most speech processing systems used to convert speech into text are tied to a single language.

Speaking rate, the number of words spoken per unit time, is a critical non-linguistic feature affecting intelligibility and comprehension of speech. Additionally, they carry patterns that are tightly correlated to the nature of the conversation. We first show a real-time implementation for monitoring speaking rate and then discuss its use in monitoring conversations effectively.

4.1 Speaking Rate

With globalization, there has been an increase in spoken interaction between people from different geographies and different cultural backgrounds. Call center or otherwise, telephone conversations are increasingly replacing face-to-face conversations

© The Author(s) 2015
S.K. Kopparapu, *Non-Linguistic Analysis of Call Center Conversations*,
SpringerBriefs in Electrical and Computer Engineering,
DOI 10.1007/978-3-319-00897-4_4

because of business needs and economics. However, unlike face-to-face conversations which are essentially driven by several unspoken cues like eye contact and body language, telephone conversations are solely driven by the cues associated with audible speech. The main reason for ineffectiveness of tele-conversations is because of the *way* speech is spoken and not necessarily by *what* is being spoken.

There is a significant influence of the background and culture of a person which determines the way the person speaks. The ineffectiveness of telephone conversation, as is common in call centers, is prominent when people from different geographies or cultures converse in a language common to them (for example English spoken in different geographies). The ineffectiveness is very prominent in a call center scenario, where the call center agent and the customer could be from two different geographies.

Speaking rate is a critical factor affecting intelligibility and comprehension of speech [1]. However, it is well known [2–5] that speaking rate varies across native and non-native speakers and additionally affects comprehension of speech in native and non-native listeners. This suggests that it is important that speakers converse at an optimal speaking rate for effective conversation. Listeners can be lost to boredom, lost to complexity, or fully engaged in a conversation based on the speakers' ability to deliver at the optimal rate [6]. The average speaking rate varies across individual speakers of the same language [7, 8].

> The average English speaking rate is between 130 and 200 *words per minute (wpm)* [9] *and applies to* 90 % *of the English-speaking population.*
>
> For complex speech content a speaking rate of 130–145 wpm is good, while 145 and 175 is considered good for speech content that is of average complexity, and for simple content, many listeners can accommodate over 175 wpm.

Good public speakers and communicators are aware of the influence of speaking rate on speech comprehension and they continuously monitor their speaking rate. They consciously adjust their conversational pace to get the message across effectively and efficiently. However, monitoring and maintaining the speaking rate at the desired levels may be hard for an average person who is

1. not aware and conscious of his speaking rate or
2. in an emotional state that does not allow him to concentrate on the speaking rate.

This emotional state influence is very prominent in call center conversations, especially for a voice agent when a customer is upset before (due to long wait time in the queue before he can speak to the agent or because of bad service provided by the enterprise) or during the conversation (because of ignorance of conversational skills or cultural knowledge of the agent).

Any platform that enables automatic monitoring of speaking rate can help speakers speak at the right speed to make conversations effective. In [1], a server-based speaking rate monitoring system that assists call center agents, in real time, to maintain an optimal speaking rate was proposed and [10] proposed a mobile phone resident application to assist a person to converse at the right speed in real-time. The essential difference being that a computationally simple algorithm is required on a mobile phone to compute the speaker rate, while one can be immune to computational complexity on a work station. We now discuss the essentials of computing speaking rate in real-time.

4.2 Speaking Rate Monitoring

In [1], a real-time speaking rate monitor that assists call center agents to maintain an optimal speaking rate in real-time was proposed. They used the number of syllables detected in speech as the measure to compute the speaking rate. The algorithm described in [11], modified to work for real-time, was used to detect syllable nuclei in spoken speech. The syllables in spoken speech were detected in the following steps:

1. All intensity peaks in the speech signal that are preceded and followed by dips in intensity are marked as potential syllables.
2. Of these, only those intensity peaks that are above a certain intensity threshold are retained as potential intensity peaks.
3. Intensity peaks that are in unvoiced region are discarded while retaining only the intensity peaks in the voiced region.
4. These intensity peaks correspond to the syllable centers of importance.

As can be noted in the third step, it is required to identify the voiced and unvoiced regions in the spoken speech to filter out the local intensity peaks in unvoiced regions. These regions are identified using the pitch contour as discussed in [12]. Once the speaking rate is computed in terms of number of syllables per second (sps) of spoken speech, we can compute the speaking rate in wpm using a conversion factor γ, as

$$S_{\text{wpm}} = \left(\frac{S_{\text{sps}}}{\gamma} \right) \times 60 \qquad (4.1)$$

where S_{wpm} is the speaking rate in words per minute, S_{sps} is the number of syllables per second, and γ is a constant that captures the average number of syllables per word; this depends on the spoken language [13].

For English language $\gamma = 1.5$ as suggested in [14].

It must be noted that since the measure is based on the number of syllables occurance in the speech, the same technique can be applied across different languages.

4.3 Speaking Rate Monitoring on Mobile Phone

As mentioned earlier, in the speaking rate monitoring system [1], the syllable detection in spoken speech happens in three essential steps. A 5 s speech window is analyzed every 1 s to enable real time monitoring of speaking rate (see Fig. 4.1).

In order to enable the real-time operation on a mobile phone, we need to modify the speaking rate computation process to work fast especially in identifying the voiced and unvoiced regions. The main computational load is due to the calculation of pitch contour, which in turn is used to distinguish voiced and unvoiced regions (Step 3). To make it real-time on a mobile phone the voiced and unvoiced detection needs to be faster on a relatively lower computing power. There are several algorithms in the literature for pitch detection [15–18] and also for identifying voiced and unvoiced regions in speech [19–21]. We used the zero-crossing rate (ZCR) to identify voiced and unvoiced regions in speech. ZCR is commonly used for voiced–unvoiced detection when the speech is clean, which is true in a call center scenario especially for the agent speech.

For purpose of illustration a clean speech sample of 5 s duration, recorded at a sampling rate of 16 kHz is used. The intensity, pitch, and ZCR for this speech sample is shown in Fig. 4.2. Intensity and pitch are calculated as in [11] and ZCR is calculated using

$$\text{ZCR} = \frac{f_\text{s}}{N} \left(\sum_{n=1}^{N-1} |\text{sgn}(x[n]) - \text{sgn}(x[n-1])| \right) \qquad (4.2)$$

where ZCR is the Zero-Crossing Rate computed as number of zero crossings per second, f_s is the sampling frequency of speech, and N is the number of speech samples in the processing window used for ZCR calculation and

$$sgn(x) = +1 \quad \text{when} \quad x \geq 0 \qquad (4.3)$$
$$= -1 \quad \text{when} \quad x < 0.$$

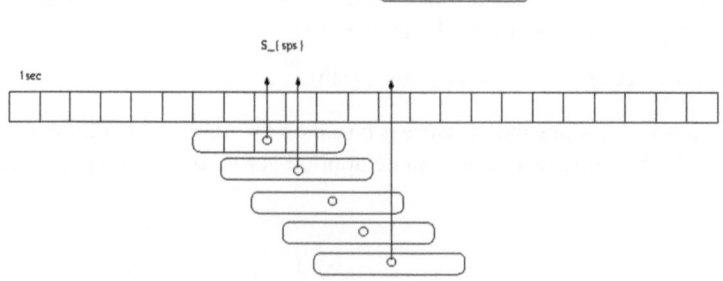

Fig. 4.1 Real-time speaking rate computing

Fig. 4.2 Syllable detection **a** using pitch and **b** using ZCR to find voiced intensity peaks

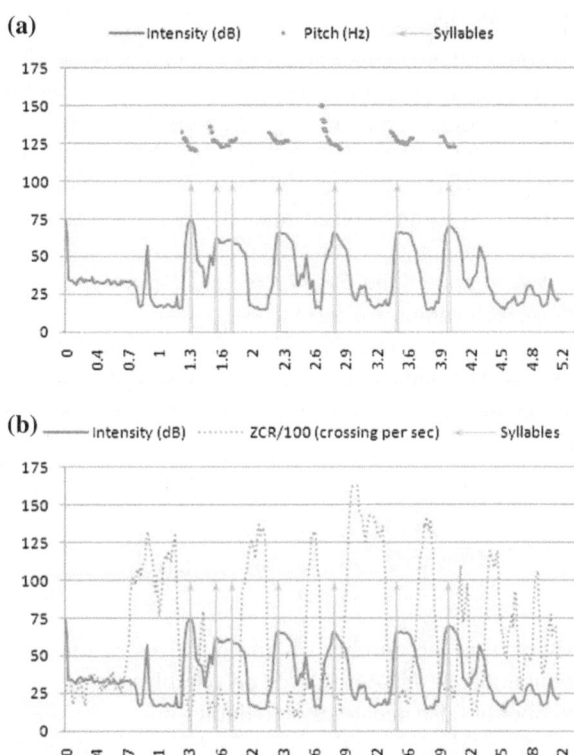

An average ZCR (>4,900 crossings per second) accounts for unvoiced speech while an average low ZCR (<1,400 crossings per second) corresponds to a voiced speech.

However, a combination of intensity and ZCR is best used to distinguish voiced and unvoiced regions in speech when the ZCR is not close to the average ZCR corresponding to voiced and unvoiced regions.

Note that both pitch contour and ZCR are used to detect voiced intensity peaks. While pitch contour extraction is computationally expensive, ZCR is not as much.

Figure 4.2 shows a comparison of syllable detection using pitch contour and ZCR to identify voiced intensity peaks. Figure 4.2a shows the intensity and pitch while Fig. 4.2b shows the intensity and ZCR for the same speech sample. It can be clearly seen (Fig. 4.2) that ZCR can be used reliably to detect voiced intensity peaks reliably. Also, as evident from (4.2), ZCR is computationally less expensive and hence can be used to calculate voiced intensity peaks (or syllables) on a mobile phone in real-time. The speed-up in the computation (using ZCR instead of pitch) comes at a small cost of accuracy of unvoiced–voiced region detection [22].

Use of ZCR to find voiced intensity peaks for syllable detection reduces the computational load and enables real-time speaking rate monitoring on the mobile phone [22]. The accuracy of syllable detection based on ZCR with syllable detection based on pitch is compared. For evaluation of accuracy (from [22]), we use speech samples from four different speakers, two male and two female, recorded using the mobile phone. The recording format is 16 bit 16 kHz PCM and the total duration of the recordings is approximately 9 min. The speakers were asked to read an English paragraph of their choice at different speaking rates. The syllable detection algorithm using pitch to find voiced intensity peaks is referred to as syl_{pitch} and the syllable detection using ZCR is referred as syl_{ZCR}.

We computed the *Precision* and *Recall* of syl_{ZCR}. Syllables are detected in all the speech samples using the syl_{pitch} and these syllables and its time of occurrence in the speech are treated as the reference. Then, syllables for the same set of speech files are detected using syl_{ZCR} and these syllables and their time of occurrence in the speech are considered as the detection hypothesis. Precision and recall are defined as

$$\text{Precision} = \frac{\text{tp}}{\text{tp} + \text{fp}} \tag{4.4}$$

$$\text{Recall} = \frac{\text{tp}}{\text{tp} + \text{fn}} \tag{4.5}$$

where tp is the number of true positives or the correctly detected syllables, fp is the number of false positives or syllables falsely detected by syl_{ZCR}, and fn is the number of false negatives or syllables missed by the syl_{ZCR}. We obtained a precision of 0.88 while the recall was 0.87 which are sufficiently high [22]. This suggests that the ZCR-based syllable detection algorithm can be used for speaking rate monitoring without much loss in the accuracy of the speaking rate values.

Further, the number of syllables for windows of 5 s duration calculated every 1 s, for all the speech samples and computed a vector X using syl_{pitch} and Y using syl_{ZCR}. We compute the correlation coefficient r as

$$r = \frac{\sum_{i=1}^{n}(X_i - \bar{X})(Y_i - \bar{Y})}{\sqrt{\sum_{i=1}^{n}(X_i - \bar{X})^2}\sqrt{\sum_{i=1}^{n}(Y_i - \bar{Y})^2}} \tag{4.6}$$

where X_i is the number of syllables detected by syl_{pitch} for the ith 5 s speech window, Y_i is the number of syllables detected by syl_{ZCR} for the ith 5 s window, $\bar{X}$ and $\bar{Y}$ are the average number of syllables detected by syl_{pitch} and syl_{ZCR}, respectively, for a 5 s speech window. The value of this Pearson's correlation coefficient r for the number of syllables detected by the two algorithms is 0.84 which indicates a high correlation between the two algorithms.

A functional real-time speaking rate measuring utility is available for download for Android phone on Google Play store [23]. *A snapshot of the utility is shown in Fig. 4.3.*

As will be shown later, speaking rate patterns do display abnormalities in the call conversations. For instance, a fast speaking customer would signify, more often,

Fig. 4.3 Monitoring speaking
rate in real-time on a mobile
phone

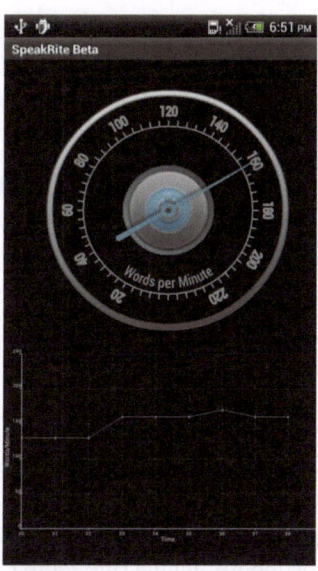

an irate customer. However, it is also important to gauge the emotional state of the
speaker in the conversation to understand the nature of audio conversation better.

4.4 Emotion in Speech

Emotion is an important non-linguistic feature that is embedded in spoken speech.
While there is a significant amount of work reported in the literature [24–27]; almost
all of them either work on (a) multiple cues, mostly speech and facial expressions
[28–30] or (b) on acted simulated speech or the same speaker emoting different
emotions explicitly [31]. Using speech as the only cue to identify the emotional state
of the speaker is research in progress, especially for non-acted or simulated speech.

In a call center voice conversation scenario, speech is the only cue that is available
for determining the emotional state of the customer or the agent. Determining the
emotional state of the customer helps in taking corrective steps by the agent to retain
the customer, especially in terms of pacifying an irate customer. While there are
several shades of observational emotions expressed by a human, from the call center
perspective it is sufficient to be able to gauge a couple of emotions like anger ($I\!R$),
neutral ($I\!N$), sad ($I\!D$), and happy ($I\!H$). Let $\mathcal{E} = \{I\!R, I\!N, I\!D, I\!H, \phi\}$ where ϕ is a label
that corresponds to no emotion state being explicitly identified.

A method based on some basic speech features, namely pitch standard deviation
(σ_p), pitch mean slope ($\bar{p}_s$), pitch average ($\bar{p}$), pitch first quantile (p_q), jitter (J),
intensity standard deviation (σ_i) has been suggested in [32]; further a rule set is also
given to determine the identified emotion [32]. However, the *if-then-else* set of rules,

Fig. 4.4 Emotion labeled for a sample speech conversation. Notice that most of the speech signals are not labeled (*white blocks*)

lands up in a zone of no emotion detected, namely the speech x is assigned the label ϕ most often. The reason for this is that the quantity of emotion displayed in the features ($\mathscr{F} = \{\sigma_p, \bar{p}_s, \bar{p}, p_q, J, \sigma_i\}$) for normal speech (non-acted, non-simulated) is not very discriminative.

See Fig. 4.4 which displays the emotion extracted from a portion of a speech conversation. The speech is first segmented into frames of 5 s and then emotion is determined for each of these frames by first extracting the feature set $\mathscr{F}$ for each frame and then subjecting it to a set of rules to bucket them into one of the emotions in $\mathcal{E}$. The labels are Red for Angry, Yellow for Happy, Green for Neutral, and Blue for Sad, clearly a large portion of the speech is assigned the label ϕ (white).

A method to associate a degree of emotion in addition to identifying the emotion in spoken speech using the speech features $\mathscr{F}$ derived from speech x can be enabled by modifying the rule set to compute $P(\bullet/x)$, the probability of the emotion $\bullet$ given the speech signal x using (4.7)–(4.9).

$$P(I\!N/x) = \frac{1}{2\pi\sigma^2}\left(w_1 e^{-\frac{(\sigma_p{}^2 - v_1)^2}{2\sigma^2}} + w_2 e^{-\frac{(\sigma_p{}^2 - v_2)^2}{2\sigma^2}}\right) * \left(w_3 e^{-\frac{(\bar{p}_s - v_3)^2}{2\sigma^2}}\right) \quad (4.7)$$

$$P(I\!H/x) = \frac{1}{2\pi\sigma^2}\left(w_1 e^{-\frac{(\sigma_p{}^2 - v_5)^2}{2\sigma^2}} + w_4 e^{-\frac{(\sigma_p{}^2 - v_6)^2}{2\sigma^2}}\right) * \left(w_1 e^{-\frac{(\bar{p}_s - v_4)^2}{2\sigma^2}}\right) \quad (4.8)$$

$$P(I\!R/x) = \frac{1}{2\pi\sigma^2}\left(w_3 e^{-\frac{(\sigma_p{}^2 - v_5)^2}{2\sigma^2}}\right) * \left(w_5 e^{-\frac{(\bar{p}_s - v_7)^2}{2\sigma^2}} + w_3 e^{-\frac{(\bar{p}_s - v_4)^2}{2\sigma^2}}\right) \quad (4.9)$$

$$P(I\!D/x) = \frac{1}{(2\pi\sigma^2)^{\frac{5}{2}}}\left(w_3 e^{-\frac{(\sigma_p{}^2 - v_1)^2}{2\sigma^2}}\right) * \left(w_3 e^{-\frac{(\sigma_i{}^2 - v_9)^2}{2\sigma^2}}\right) \quad (4.10)$$
$$* \left(w_7 e^{-\frac{(J - v_{10})^2}{2\sigma^2}} + w_1 e^{-\frac{(J - v_{11})^2}{2\sigma^2}}\right)$$
$$* \left(e^{-\frac{(\bar{p} - v_3)^2}{2\sigma^2}} * \left\{w_6 e^{-\frac{(p_q - v_{12})^2}{2\sigma^2}} + w_5 e^{-\frac{(p_q - v_{13})^2}{2\sigma^2}}\right\}\right.$$
$$\left. + e^{-\frac{(\bar{p} - v_3)^2}{2\sigma^2}} * \left\{w_5 e^{-\frac{(p_q - v_{14})^2}{2\sigma^2}} + w_6 e^{-\frac{(p_q - v_{15})^2}{2\sigma^2}}\right\}\right)$$

Fig. 4.5 Emotion labeled for a sample speech conversation

The emotion with the highest probability, namely

$$\max\{P(I\!N/x),\, P(I\!H/x),\, P(I\!R/x),\, P(I\!D/x)\}$$

determines the emotion of the speech x. Note that all the weights w_k and constants v_k in (4.7), (4.8), (4.9), and (4.10) are derived from [32]. Figure 4.5 displays the emotion extracted for the same portion of the speech conversation shown in Fig. 4.4. Clearly, the entire speech signal is assigned a valid emotion label. Observe that there are no speech frames that have been labeled ϕ.

4.5 Conclusion

Speaking rate and emotion are non-linguistic characteristics of speech that are able to determine how the speech was spoken. Speaking rate affects the intelligibility and comprehension of speech, especially during telephone conversations when there are no clues except audio; this feature is able to convey a lot of hidden information in terms of conversational quality. Automatic extraction of speaking rate and emotion can help to understand the conversation and can be effectively used to monitor call center conversations.

References

1. M.A. Pandharipande, S.K. Kopparapu, Real time speaking rate monitoring system. in *International Conference on Signal Processing, Communications and Computing (ICSPCC)*, Sept. 2011, pp. 1–4
2. A. Janet, K. Kenneth, The effect of foreign accent and speaking rate on native speaker comprehension. Lang. Learn. **38**, 561–613 (1988)
3. J. Murray, M. Tracey, The effects of speaking rate on listener evaluations of native and foreign-accented speech. Lang. Learn. **48**, 159–182 (1998)
4. J. Yuan, M. Liberman, C. Cieri, Towards an integrated understanding of speaking rate in conversation. Proc. Interspeech **2006**, 541–544 (2006)
5. A. Bradlow, D. Pisoni, Recognition of spoken words by native and non-native listeners: talker-, listener-, and item-related factors. J. Acoust. Soc. Am. **106**(4), 2074–2085 (1999)
6. D. O' Sullivan, Caller adaptive voice response system. US Patent 5,493,608, 20 February 1996
7. M.J. Munro, T.M. Derwing, The effects of speaking rate on listener evaluations of native and foreign-accented speech. Lang. Learn. **48**(2), 159–182 (1998)
8. M.J. Munro, T.M. Derwing, Modeling perceptions of the accentedness and comprehensibility of l2 speech the role of speaking rate. Stud. Second Lang. Acquis. **23**, 451–468 (2001)

9. D O' Sullivan, Using an adaptive voice user interface to gain efficiencies in automated calls. Whitepaper, Interactive Digital, (2009)
10. I. Ahmed, M. Pandharipande, S.K. Kopparapu, *Speakrite: A Real Time Tool for Speaking Rate Monitoring.* in SiMPE, (2012)
11. N.H. De Jong, T. Wempe, Praat script to detect syllable nuclei and measure speech rate automatically. Behav. Res. Methods **41**, 385–390 (2009)
12. P. Boersma, *Accurate Short-term Analysis of the Fundamental Frequency and the Harmonics-to-Noise Ratio of a Sampled Sound.* in institute of phonetic sciences, University of Amsterdam, Proceedings 17, (1993)
13. P. Francois, C. Christophe, M. Egidio, Across-language perspective on speech information rate. Language, Sept. 2011, pp. 539–558
14. J. Scott, Yaruss. Converting between word and syllable counts in children's conversational speech samples. J. Fluency Disord. **25**(4), 305–316 (2000)
15. P. Vepreka, M. Scordilis, Analysis, enhancement and evaluation of five pitch determination techniques. Speech Commun. **37**, 249–270 (2002)
16. A. Camacho, *Swipe: a sawtooth waveform inspired pitch estimator for speech and music.* PhD thesis, University of Florida, Gainesville, FL, USA, 2007, AAI3300722
17. M.G. Christensen, A. Jakobsson, Multi-pitch Estimation. *Synthesis Lectures on Speech and Audio Processing.* Morganand Claypool Publishers, (2009)
18. P.N. Garner, M. Cernak, P. Motlicek, A simple continuous pitch estimation algorithm. IEEE Signal Process. Lett. **20**(1), 102–105 (2013)
19. M. Radmard, M. Hadavi, M. Nayebi, A new method of voiced unvoiced classification based on clustering. J. Signal Inform. Process. **2**, 336–347 (2011)
20. K.I. Molla, K. Hirose, N. Minematsu, and K. Hasan. Voiced/unvoiced detection of speech signals using empirical mode decomposition model. in *International Conference on Information and Communication Technology, ICICT '07*, March 2007, pp. 311–314
21. R.G. Bachu, S. Kopparthi, B. Adapa, B.D. Barkana. Voiced/Unvoiced Decision for Speech Signals Based on Zero-Crossing Rate and Energy. in Khaled Elleithy, editor, Advanced Techniques in Computing Sciences and Software Engineering, Springer, Netherlands, 2010, pp. 279–282
22. I. Ahmed, M. Pandharipande, S.K. Kopparapu, Speakrite: monitoring speaking rate on mobile phone in real time. Int. J. Mobile Human Comput. Interac. **5**(1), 62–69 (2013)
23. YP. Awaz, Speakrite, 2013, https://play.google.com/store/apps/details?id=com.tcs.android speech.speedometer.main,
24. I. Luengo, E. Navas, I. Hernáez, Feature analysis and evaluation for automatic emotion identification in speech. IEEE Trans. Multimedia **12**(6), 490–501 (2010)
25. M. Shah, L. Miao, C. Chakrabarti, A. Spanias, A speech emotion recognition framework based on latent dirichlet allocation: algorithm and fpga implementation. in *IEEE International Conference on Acoustics, Speech and Signal Processing (ICASSP)*, May 2013, pp. 2553–2557
26. B. Schuller, A. Batliner, S. Steidl, D. Seppi, ecognising realistic emotions and affect in speech: state of the art and lessons learnt from the first challenge. Speech Commun. Sens. Emot. Affect - Facing Realism Speech Process. **53**(910), 1062–1087 (2011)
27. I. Lefter, Automatic emotion enalysis based on speech, (2009)
28. C. Busso, Z. Deng, S. Yildirim, M. Bulut, C.M. Lee, A. Kazemzadeh, S.Lee, U. Neumann, S. Narayanan, Analysis of emotion recognition using facial expressions, speech and multimodal information. in *Proceedings of the 6th International Conference on Multimodal Interfaces*, ICMI '04, New York, NY, USA, 2004, pp. 205–211
29. R. Lpez-Czar, J. Silovsky, M. Kroul, Enhancement of emotion detection in spoken dialogue systems by combining several information sources. Speech Commun. Sens. Emot. Affect - Facing Realism Speech Process. **53**(910), 1210—1228 (2011)
30. I. Stankovic, M. Karnjanadecha, V. Delic, Improvement of thai speech emotion recognition by using face feature analysis. in *International Symposium on Intelligent Signal Processing and Communications Systems (ISPACS)*, Dec 2011, pp. 1–5

31. T.-L. Pao, C.-H. Wang, and Y.-J. Li, A study on the search of the most discriminative speech features in the speaker dependent speech emotion recognition. in *Fifth International Symposium on Parallel Architectures, Algorithms and Programming (PAAP)*, Dec 2012, pp. 157–162
32. D.B. Ryan, 2003, http://www.denisryan.com/thesis/capstoneV1.02.pdf

Chapter 5
Case Study

5.1 Introduction

Voice-based call centers enable customers to query for information by speaking to human agents. Most often these call conversations are recorded by call centers with the intent of trying to identify things that can help improve the performance of the call center to serve the customer better. Today, recorded conversations are analyzed by humans by listening to call conversations, which is both time-consuming, fatigue prone, and not very accurate. Additionally, humans are able to analyze only a small percentage of the total calls because of economics.

However, with the advent of automatic speech recognition (ASR) technology, the task of identifying problematic calls reduces to one of automatic transcription of telephone calls followed by analyzing the words or phrases in the text transcriptions [1]. However, the process of automatically converting audio conversations into transcribed text is not very accurate, typically the recognition accuracies are around 50–60 % even if one uses the state of the art speech recognition technology. The recognition accuracies further deteriorate when there is no readily available ASR for the language spoken in the conversation [2]. In many cases, transcription of the audio conversation is not sufficient to identify a problem situation in a call conversation, because the problem in the call might not translate into meaningful transcribed text. For example, /thank you/ spoken in a sarcastic tone might not suggest a problem in the call because the phrase "Thank You" is generally associated with a satisfied customer, and hence would be categorized as a non-problematic call.

In this chapter, (based on [3]), a visual method to enable automatic identification of problematic calls without actually transcribing the audio conversations into text is discussed. The idea is to sieve through *all* the calls and identify problem calls; these calls can then be further analyzed by humans if desired or required. We first model call conversations as a directed graph (see Fig. 3.1) using speaking rate [4] feature to abstract the call conversation and then identify a directed graph structure associated with a normal call. We use the speaking rate [4] feature to abstract the call conversation and use directed graph to represent a call conversation. All call

© The Author(s) 2015
S.K. Kopparapu, *Non-Linguistic Analysis of Call Center Conversations*,
SpringerBriefs in Electrical and Computer Engineering,
DOI 10.1007/978-3-319-00897-4_5

conversations that do not have the structure of a normal call are then classified as being abnormal or problematic. We use the speaking rate feature to model call conversation because it can spot potential problem calls. We have experimented on real call center conversations acquired from different call centers.

5.2 Related Work

Identifying problematic call conversations helps the call center improve its performance in terms of customer satisfaction, customer retention, cross-selling, and process improvement to name a few. A spotted problematic call gives a lot of insight into possible process and people improvements that can enhance the performance of a call center. In addition to these benefits, the ability of the supervisor to efficiently pinpoint personalized training to the call center voice agents is very important.

There are two main approaches that call centers follow to identify problematic calls. The first approach is based on looking at the meta data associated with the call, namely if the call takes a longer time than usual to complete, or if the call is transferred to a supervisor for completion, then it is flagged as being problematic. As an extension, in case call centers are equipped with multiple interaction channels, then a complaint from a customer in the form of an email is associated with the voice call to mark it as being problematic. These approaches are not reliable as they look only at the meta data associated with the voice call and not the actual call conversation itself. Very often these meta data-based approaches miss out on important problematic calls that do not leave any cues in the meta data.

The second approach, as explained earlier, is to transcribe the call conversation using an automatic speech recognition engine and then sieve through the text transcriptions to flag calls based on *key* words or phrases. This approach has several drawbacks, the first and the foremost is the fact that the state of the art call center conversation transcription accuracy is very noisy and erroneous, and the second important factor is that there are several instances, as described earlier, when just analyzing the text transcription does not yield clues that can be associated with a call being labeled as being normal or problematic.

Hironori [5] describes a method to identify important segments from transcribed textual records of conversations between customers and agents. They look for changes in the accuracy of a categorizer designed to separate different business outcomes.

Gilad [6] describes a system that automatically transcribes calls using a speech recognition engine. The domain-specific importance of the conversation fragments is identified based on the divergence of corpus statistics. This is used to analyze the content of the call conversation. They further use information retrieval approaches on the transcribed text to provide knowledge mining tools for both call-center agents and for administrators of the center. The system developed in [6] helps in gaining insight hidden in the recorded calls, which can help reduce the cost of operation and improve products, processes. This enables making quality monitoring more effective by routing calls on key business issues to the supervisor for review affecting the overall customer experience.

Vincenzo [7] provides a solution for pragmatic analysis of call center conversations in order to provide useful insights for enhancing Call Center Analytics to a level that will enable new metrics and Key Performance Indicators (KPIs) beyond the standard approach. These metrics rely on understanding the dynamics of conversations by highlighting the way participants discuss about topics. By this, they claim, one can detect situations that are simply impossible to detect with standard approaches such as controversial topics, customer-oriented behaviors, and also predict customer ratings.

In summary, most of the work described in the literature is carried out on the transcribed call conversation [6, 8, 9], which means we have to depend on the not so accurate ASR for audio transcribed text. However, there are several languages, generally called the resource deficit language, that do not have even a basic speech-to-ext conversion engine. Even for languages like English, where language resources exist and several ASRs exist, the typical speech to text conversion yields only 50–60 % recognition accuracies (see Appendix F).

We propose a method to flag problematic calls by neither looking at the meta data associated with the call nor requiring us to go through the poor accuracy speech to text transcription. We use

1. the ability to distinguish the call conversation into segments spoken by the agent and those spoken by the customer and
2. non-linguistic feature associated with the call conversation, namely speaking rate and emotion

to identify and flag problematic calls. Clearly, this approach is independent of the language of communication in the call conversation.

5.3 Modeling Call Conversations

A typical call center conversation between the agent in the call center and the customer is a sequence of speech segments spoken by either the agent or the customer. Generally,the call is initiated by the customer and lands on the interactive voice response (IVR) system. The call is then routed to one of the several agents who are free at that time to receive a call; the voice agent then starts the conversation. A call center conversation can be represented as a directed graph as shown in Fig. 5.1. The nodes represent the person (agent or customer) who is speaking and edges represent the transition between the speakers.

Typically, in almost all call centers, at the beginning of a call, the agent starts with a welcome message, and it is the agent who ends the call with a goodbye or thank you message. As seen in Fig. 5.1 the red arrow into the AGENT SPEAK node shows the start of the conversation, while the blue arrow, going out of the AGENT SPEAK

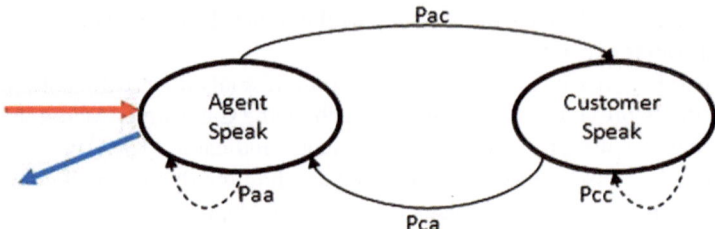

Fig. 5.1 Directed graph representation of a call center conversation

node shows the end of the conversation. The rest of the conversation is spoken by the customer (CUSTOMER SPEAK) or the agent (AGENT SPEAK) and is interlaced, this is shown by the continuous black lines in Fig. 5.1. The dashed lines in Fig. 5.1 represent the agent (customer) speaking continuously with possible pauses without allowing the customer (agent) to speak. The edge weights P_{cc}, P_{ca}, P_{ac} and P_{aa} represent the probability that the customer continues to speak with possible pauses, probability that the agent speaks after the customer, probability that the customer speaks after the agent and probability that the agent continues to speak with possible pauses, respectively.

In some sense Fig. 5.1 gives an abstract view of a typical call center conversation. The size of the node captures the total talk time of that speaker during the call. For example, a large AGENT SPEAK compared to CUSTOMER SPEAK node means that the agent was conversing more during the call. Additionally, the edges in the graph determine and capture the nature of the conversation. For example a missing edge going out of the AGENT SPEAK node at the end of the conversation is likely to be a problematic call or if $P_{cc} \gg P_{ca}$ then the customer is speaking more and, possibly, not allowing the agent to speak or respond to his problem; in majority of the cases this might translate to a problematic call.

> During a normal interaction between the customer and the agent one expects $P_{ac} > P_{aa}$ and $P_{ca} > P_{cc}$ where the agent speaks and also allows the customer to speak and vice-versa.

However, a realistic call conversation has *hold music*, the time during the conversation, when neither the agent nor the customer is speaking; this happens when the agent is fetching the information sought by the customer from an IT backend system, so there is another node, namely HOLD MUSIC associated with the conversation.

> In a typical conversation the HOLD MUSIC state is entered from the AGENT SPEAK state and exits to AGENT SPEAK state.

Clearly, if the conversation stays longer in the HOLD MUSIC state (large P_{mm}, see Fig. 5.2.), then it can be assumed that the agent is not performing efficiently because he is putting the customer on wait for a longer duration or the customer has a complex query that the agent is not being able to address. Along similar lines, if P_{am} is larger than P_{aa} or P_{ac} then it is very likely that the agent is putting the customer on hold a large number of times indicating that the agent is unable to resolve the customer

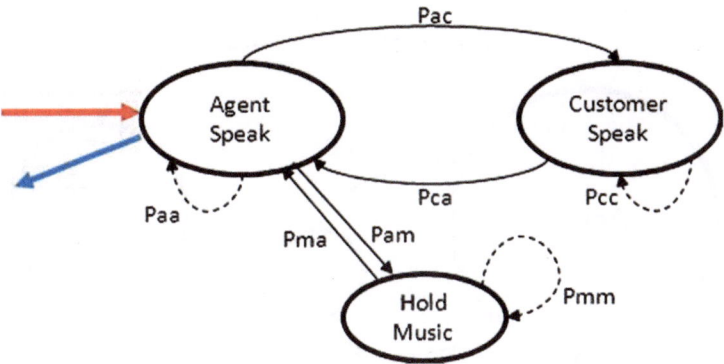

Fig. 5.2 Directed graph representation of a realistic call center conversation

query. These situations are reasonably good indicators of a problem situation in a call center conversation. Clearly, this mode of modeling a call center conversation enables one to understand certain aspects of the call in terms of performance of the agent. Understandably, both (a) the probabilities P_{**} and (b) the size of the nodes, AGENT SPEAK, HOLD MUSIC and CUSTOMER SPEAK are computed on the complete call, and hence gives an overall picture of the call.

The probabilities P_{**} and the size of the nodes, in a call conversation are computed as shown in Appendix C.

The probabilities P_{**} do not in themselves help in identifying the portions of the call that might not be normal.

As will be shown later, the directed graph representation of the call conversation has a different structural pattern, which depends on the type of transaction happening during the call.

To check the use of the directed graph to identify abnormal calls, we randomly selected 75 actual call center conversations that we had access to from different call centers. Figure 5.3 shows a typical directed graph of a normal call, while Fig. 5.4 shows the directed graph of an abnormal or a problematic call. As a first step the calls were automatically segmented into voice and music [10]. Further, these voice segments were segmented into sections spoken by agent and customer using an automated method described in [11]. Subsequently, we found the number of transitions from one node (here node refers to AGENT SPEAK, CUSTOMER SPEAK and HOLD MUSIC) to another and also calculated the duration of speech in a particular node (as described in Appendix C). It was found that in a normal call the probability of agent talking to customer P_{ac} is lesser than the probability of customer talking after an agent has spoken P_{ca} ($P_{ac} < P_{ca}$).

Also, the probability of an agent putting the customer on hold P_{am} is much lesser than the probability of agent talking to customer P_{ac}, namely $P_{am} \ll P_{ac}$. Note that the size of the AGENT SPEAK node (P_{aa}) is much larger than the node corresponding to CUSTOMER SPEAK (captured by P_{cc}) meaning the agent is speaking for a longer

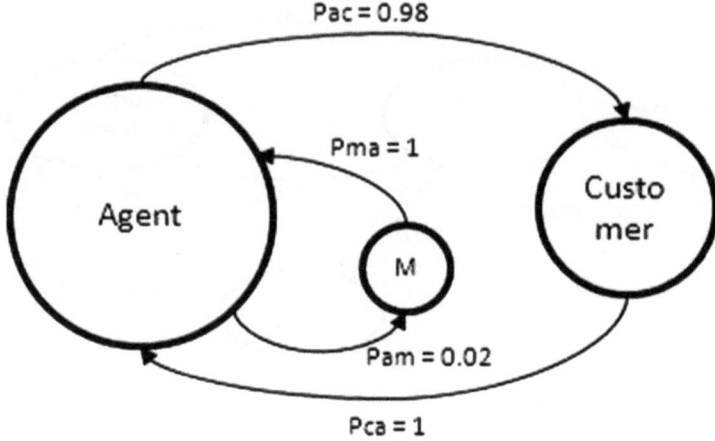

Fig. 5.3 Directed graph representation of an actual normal call

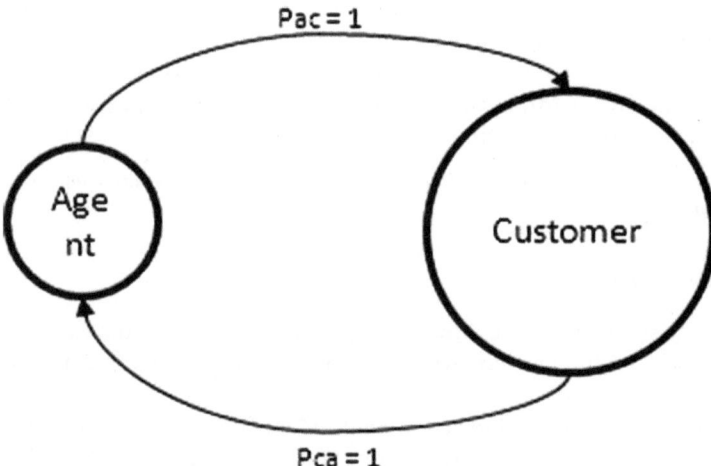

Fig. 5.4 Directed graph representation on an actual abnormal call

duration than the customer. This in general symbolizes a typical normal call. While in an abnormal call see Fig. 5.4 the size of the node corresponding to the CUSTOMER SPEAK is much bigger than the size of the node corresponding to AGENT SPEAK typically indicating a problematic call.

However, this analysis based on directed graph is not sufficient to identify the actual location of the abnormality, if it exists, in a call. We need to model the call at a better time resolution, and we discuss this next where we use the speaking rate as an additional non-linguistic feature to represent the call conversation.

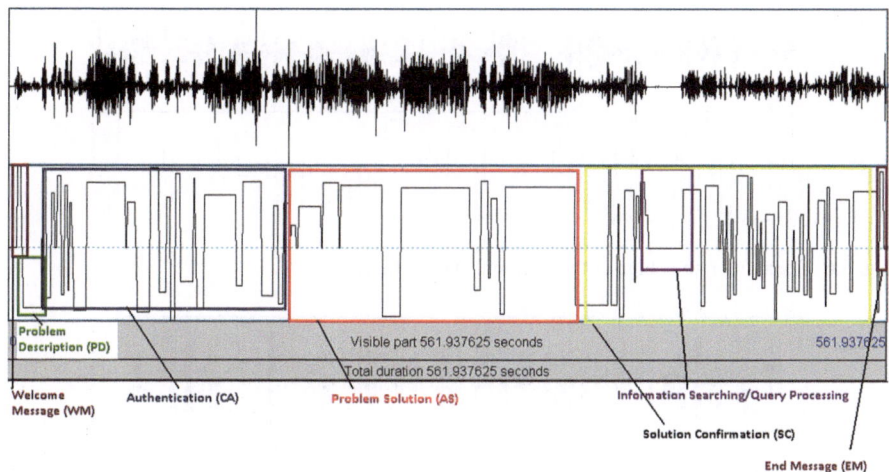

Fig. 5.5 Different transactions in a call conversation and the associated speaking rate pattern

5.4 Speaking Rate Feature-Based Analysis

For analysis of call conversation we use the speaking rate feature (see Appendix B). The speaking rate is computed on a segment of the call conversation spoken by the same person (agent or customer) in one stretch. As seen in Fig. 5.5, a typical call conversation can be segmented into portions of speech spoken by the agent (AGENT SPEAK), that spoken by the customer (CUSTOMER SPEAK), and the hold music (HOLD MUSIC). Each of this segment, which is not hold music, can be analyzed to estimate the speaking rate (S_{wpm}), and the speaking rate during hold music segment is assumed to be zero. The speaking rate of the agent is shown on the positive y-axis, and that of the customer is shown on the negative y-axis. The amplitude represents the speaking rate; larger amplitude represents a higher speaking rate and vice versa A typical speaking rate pattern of a complete call conversation of 598.7 s duration is shown in Fig. 5.5.

> *Note the speaking rate on the negative y-axis for the customer is for representational purpose only. One needs to consider the absolute value of the speaking rate; a higher absolute value would represent faster speaking rate.*

5.5 Speaking Rate Patterns in Call Conversations

There are several patterns that one can observe in call conversations that are typical of a conversation between the customer and the agent. Speaking rate (S_{wpm}) of the agent is high at the start of the conversation as well as at the end of a call conversation as seen in Fig. 5.6.

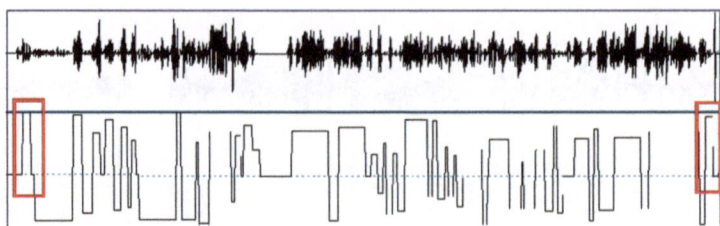

Fig. 5.6 Speaking rate is high at the start and end of the call

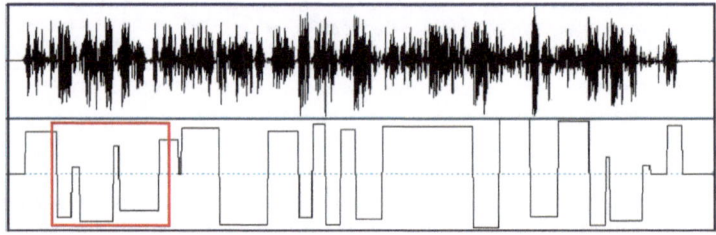

Fig. 5.7 Problem description by customer

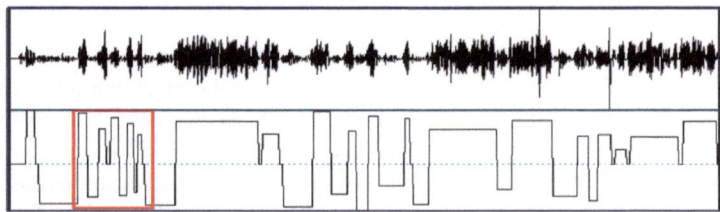

Fig. 5.8 Authentication process

This is to be expected as very often the agent is either reading a predrafted script or has spoken the paragraph so many times that it becomes his second nature.

For example: /very good evening, welcome to <company> care this is <name> how may I assist you/ at the beginning of the call or /thank you for calling <company> care have a nice day/ at the end of the call.

A customer describing the problem shows up as a pattern shown in Fig. 5.7. The pattern has more or less a constant speaking rate with intermittent agent speaks of very short duration. Figure 5.8. shows the authentication procedure. Typically, the agent makes sure that he is indeed speaking to the person whom he is supposed to speak. Asking for name or contact number or card details to verify and authenticate the customer. Agent providing a solution in response to the customer query has a pattern shown in Fig. 5.9. Typically, the agent speaks for a length of time with small pauses in between; this is an indicative pattern of the agent providing a verbal solution to the customer query and an agent searching for some information has a pattern shown in Fig. 5.10.

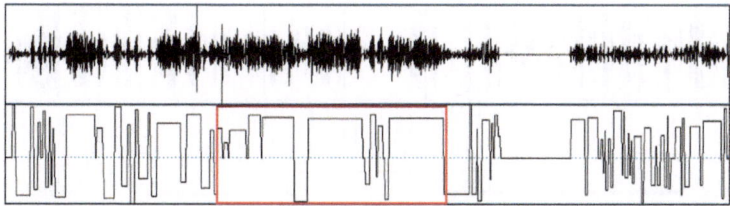

Fig. 5.9 Solution being provided by agent

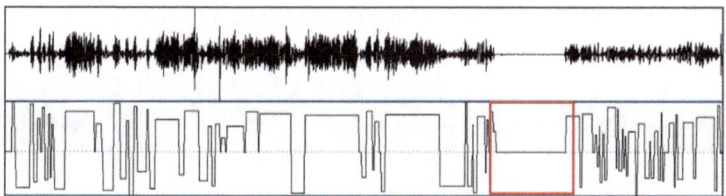

Fig. 5.10 Agent searching for information

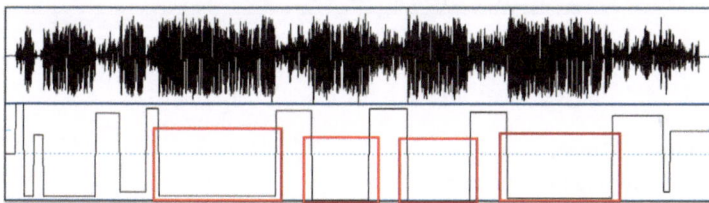

Fig. 5.11 Abnormal call: customer not allowing agent to speak

We can observe that the speaking rate feature displays unique patterns, which capture certain aspects of the call conversation. Additionally, using speaking rate feature we can very clearly identify all those instances in the conversation where the customer's speaking rate is continuously high. Figure 5.11, for example, captures an instance in a conversation where the customer is speaking continuously with a high speaking rate, this pattern is an indication of the customer not being happy for some reason. Also conversation ending abruptly with no /Thank you/ message from the agent (Fig. 5.12) is an indication of an upset customer who hung off the call midway.

The observed patterns in the speaking rate feature can be associated with a typical transaction within the call conversation. Typical transactions in a conversation could be

1. [*WM*] Welcome Message by the agent.
2. [*PD*] Problem Description by customer.
3. [*CA*] Customer Authentication.
4. [*AS*] Agent providing a Solution.
5. [*SC*] Agent Confirming customer understanding the solution, solution confirmation.

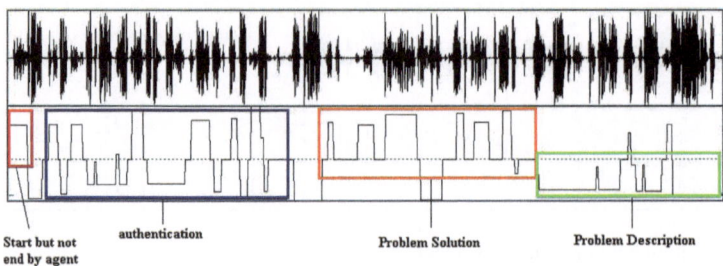

Fig. 5.12 Abnormal call: call ends abruptly no /*Thanks message*/ from agent

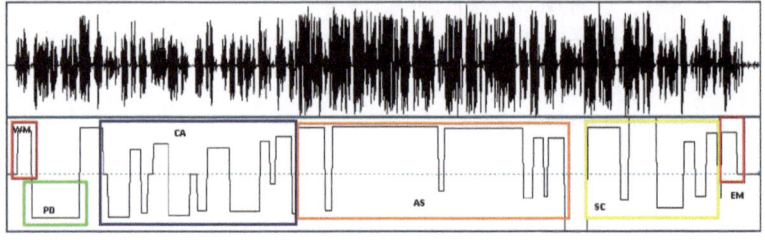

Fig. 5.13 Typical speaking rate pattern of a call conversation

6. [*OP*] Pauses, Music.
7. [*EM*] End Message by the agent.

Figure 5.13 (like Fig. 5.5) captures a complete call center conversation. The red box at the beginning is the [*WM*] component and indicates the start, and the red box at the end of call indicates the component [*EM*]. The green box indicates description of problem by customer, which is the [*PD*] component, while the blue box indicates the authentication process [*CA*], where the agent makes sure that he is indeed speaking to the person he should be speaking to. The orange box indicates a typical solution being provided by the agent to customer [*AS*]. The yellow box indicates confirmation of the solution provided by the agent and affirmation by the customer to the solution [*SC*]. Table 5.1 captures this in a nutshell. Each of these transactions within the call conversation can be represented as a directed graph, consisting of AGENT SPEAK, CUSTOMER SPEAK and HOLD MUSIC.

Figures 5.14 and 5.15 capture the structure of the directed graph in each of these transactions for two different call conversations. The y-axis shows the speaking rate and the x-axis is an indication of the time in the conversation. The size or the area of the node captures the total speaking time during the transaction and the ellipticity of the node shows the variation in the speaking rate during the transaction. Clearly, in the transaction corresponding to [WM], only the agent is speaking (larger AGENT SPEAK), and he is speaking fast (location of the AGENT SPEAK node is higher up along the y-axis).

Table 5.1 Patterns in a normal call

Transaction	Pattern (SR)	Who
Welcome message (WM)	High	Agent
Problem description (PD)	Uniform	Customer
Authentication (CA)	Low	Agent-customer
Agent solution (AS)	Uniform	Agent
Solution confirmation (SC)	Uniform	Agent-customer
End message (EM)	High	Agent

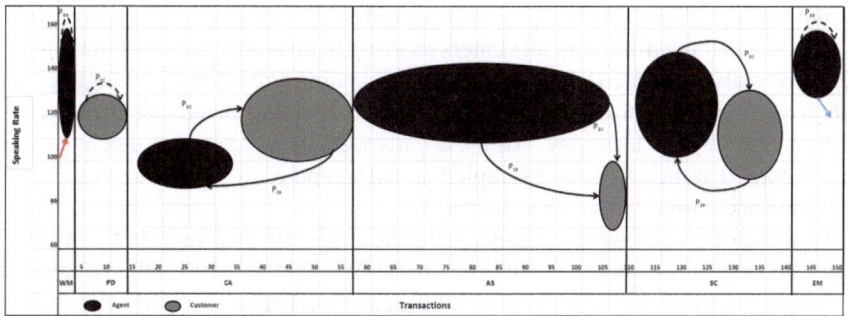

Fig. 5.14 Structure of the directed graph in each transaction

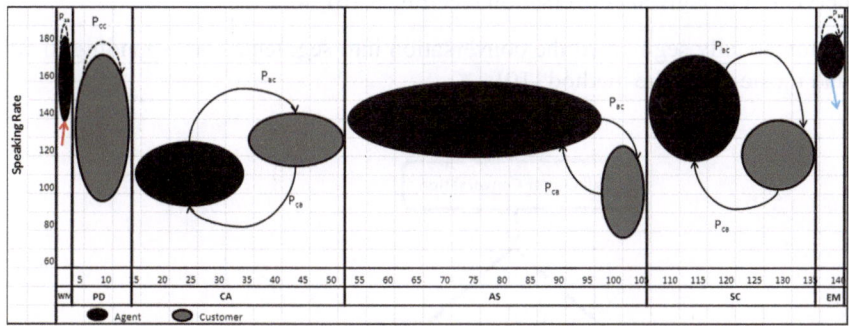

Fig. 5.15 Directed graph in each transaction for a different call

Note that the emotion feature, as described earlier, can be computed for each of the spoken segment separately, and can be presented along with the speaking rate feature, making the emotion of the speaker known in each transaction.

As seen in Fig. 5.14, a normal call has the following characteristics, namely,

1. A normal call has a very specific pattern in terms of transactions components during the length of the conversation, and
2. A normal call has typical components, which are organized in a particular time sequence.
3. Normal call conversations are structurally similar as seen in Figs. 5.14 and 5.15.

In our experiments, we used the structure defined by Figs. 5.13 and 5.14, also captured in Table 5.1, to identify a normal call. We compared test call conversation with this reference model to identify how close the call was to a normal call. Any deviation from the normal structure was flagged as being abnormal.

5.6 Experimental Results

For the purpose of experiments [3], we analyzed a set of 100 call center conversations obtained from three different call centers catering to insurance and telecom domains. While the calls associated with the insurance domain were in English language, they were from two different geographies; the telecom domain-related calls were in a mix of Hindi and English language. All these calls were actual customer-agent calls and were recorded at the call center sampled at 8 kHz, and each sample was represented by 16 bits and stored in the *.wav* format.

All these 100 calls were carefully listened to by atleast a supervisor in the call center, and each was marked as abnormal or normal. Of these 100 calls, 10 calls were marked abnormal. For example, in one of the calls marked as abnormal, the customer had put down the phone without allowing the agent to even speak his typical thank you message. The other 90 calls were marked as being normal by the call center supervisor. For a given conversation (see Fig. 5.16), we

- Automatically segmented the conversation into segments corresponding to voice and music using the method [10],

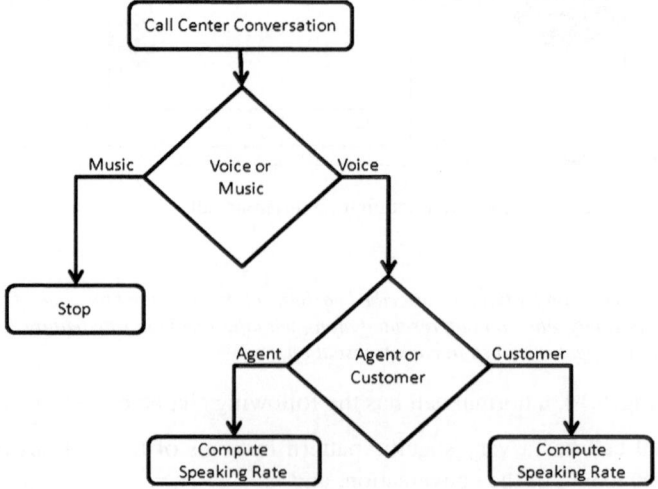

Fig. 5.16 Computing the speaking rate (and emotion) of agent and customer

- and then all the segments that were marked as voice were further automatically segmented into spoken by agent or spoken by customer refer [11].
- For each of the voice segment, we automatically computed the speaking rate of the segment using the method mentioned in [4].

We marked the conversations, semi-automatically, using the speaking rate feature to determine the type of transaction, namely one of [WM], [PD], [CA], [AS], [SC], [OP], [EM] without actually hearing to the audio conversation (using Table 5.1).

We selected, at random, a few call conversations to check the correctness of the marking of the transaction by manually listening to the audio segments, and found that we were able to mark with more than 90 % accuracy. Now, each call conversation was first segmented into voice and music [10], and then all the segments that were marked as voice were segmented into spoken by agent or spoken by customer [11]. We then constructed the directed graph for each transaction in the call conversation by first labeling the segment with one of the labels ([WM], [PD], [CA], [AS], [SC], [OP], [EM]) along with the time taken for that transaction. Any call that (a) missed one or more labels, or (b) a certain label that had an unusual duration was marked as being abnormal. This process yielded 90 % results in the sense that we were unable to mark just 1 abnormal call in the 10 abnormal calls in our dataset. However, as many as 7 normal calls were marked as abnormal because we had falsely mislabeled the conversation with transaction labels based on the observed pattern.

Non-linguistic speech features like speaking rate and emotion can be used effectively to identify problematic call conversation without actually having to explicitly know what linguistic content is being conversed. While non-linguistic analysis is able to overcome the inaccuracies of a speech-to- text conversion process, it also makes it possible to analyze conversation in any language for which a speech recognition engine is not readily available. Some experimental results for Spanish call center conversation should be available at https://sites.google.com/site/nlccanalytics/.

References

1. VERINT, Speech analytics essentials for audiolog, http://www.verint.com/solutions/enterprise-workforce-optimization/products/speech-analytics/impact-360-speech-analytics-essentials-for-audiolog
2. S.K. Kopparapu, I. Ahmed, Enabling rapid prototyping of an existing speech solution into another language, in *14th Oriental COCOSDA Conference*, Oct 2011
3. M.A. Pandharipande, S.K. Kopparapu, A novel approach to identify problematic call center conversations, in *Computer Science and Software Engineering (JCSSE), 2012 International Joint Conference on*, 30–1 June 2012, pp. 1–5
4. M.A. Pandharipande, S.K. Kopparapu, Real time speaking rate monitoring system, in *Signal Processing, Communications and Computing (ICSPCC), 2011 IEEE International Conference on*, sept. 2011, pp. 1–4
5. H. Takeuchi, L.V. Subramaniam, T. Nasukawa, S. Roy, Automatic identification of important segments and expressions for mining of business-oriented conversations at contact centers, in *Joint Conference on Empirical Methods in Natural Language Processing and Computational Natural Language Learning (EMNLP-CoNLL)*, 2007

6. G. Mishne, D. Carmel, R. Hoory, A. Roytman, A. Soffer, Automatic analysis of call-center conversations, in *Proceedings of the 14th ACM International Conference on Information and Knowledge Management, CIKM '05*, New York, USA, 2005, ACM, pp. 453–459

7. V. Pallotta, R. Delmonte, L. Vrieling, D. Walker, Interaction mining: the new frontier of call center analytics, in *5th International Workshop on New Challenges in Distributed Information Filtering and Retrieval*, 2011

8. G. Zweig, O. Siohan, G. Saon, B. Ramabhadran, D. Povey, L. Mangu, B. Kingsbury, Automated quality monitoring for call centers using speech and nlp technologies, in *Proceedings of the 2006 Conference of the North American Chapter of the Association for Computational Linguistics on Human Language Technology: Companion Volume: Demonstrations, NAACL-Demonstrations '06*, Stroudsburg, PA, USA, 2006, Association for Computational Linguistics, pp. 292–295

9. F. Cailliau, A. Cavet, Mining automatic speech transcripts for the retrieval of problematic calls, in *Proceedings of the 14th International Conference on Computational Linguistics and Intelligent Text Processing—Volume 2, CICLing'13*, Berlin, Heidelberg, 2013, Springer, pp. 83–95

10. S.K. Kopparapu, M.A. Pandharipande, G. Sita, Music and vocal separation using multiband modulation based features, in *Industrial Electronics Applications (ISIEA), 2010 IEEE Symposium on*, Oct 2010, pp. 733–737

11. S.K. Kopparapu, A. Imran, G. Sita, A two pass algorithm for speaker change detection, in *TENCON 2010–2010 IEEE Region 10 Conference*, nov. 2010, pp. 755–758

Chapter 6
Conclusions

There is a rich source of information begging to be exploited in the customer-agent voice conversations which can enhance the customer satisfaction index and other performance metrics of a voice-based call center. Additionally, information derived from the call conversation can be used to identify personalized training needs of the agent as well as quickly address areas of concern raised by specific customers. Voice-based call centers have the option of depending on individual supervisors in the call centers to manually analyze a small sample of the recorded call conversation to derive usable analytics. There are several issues associated with performing the task of analyzing call center conversations; the large amount of voice data is one of them.

However, there is a growing awareness within the call center industry to exploit the information hidden in the call conversations which has led to voice-based call centers adopting automatic and learning-based methods to analyze call conversations.

The tools that have been adopted to enable speech analytics have been restricted to the process of first converting the audio conversations into text using a speech recognition engine, followed by analyzing the noisy transcription using natural language text processing techniques. This process of speech to text followed by text analysis, though natural, is not only expensive but also erroneous because of the speech-to-text conversion process not being very accurate.

The two-step process requires a speech to text conversion, which in itself returns poor recognition accuracies even with the state-of-art speech recognition engines built for resource rich languages; the recognition accuracies are further hampered for resource deficit languages or conversation which has interplay of two or more languages. Additionally, since the analysis is based on text, there is a challenge in determining a /thank you/ said with sarcasm versus a genuine /thank you/. This poses a serious challenge in determining an abnormal or a problematic call from a normal call conversation just based on analyzing textual description of the conversation.

A novel way of using non-linguistic features to enable to identify an abnormal call from a normal call center conversation without actually converting the audio conversation into text or manually listening to the call has been motivated and discussed in this monograph.

© The Author(s) 2015
S.K. Kopparapu, *Non-Linguistic Analysis of Call Center Conversations*,
SpringerBriefs in Electrical and Computer Engineering,
DOI 10.1007/978-3-319-00897-4_6

The use of directed graph to represent a call center conversation between the agent and the customer is able to visually represent the goodness or abnormality of the conversation. Additionally, the use of a non-linguistic feature, namely speaking rate, enables identification of different types of transactions within a call conversation. The identification of the speaking rate patterns and mapping them to transactions in the call conversation enables identification of the transaction as being normal or not. The main advantage of using the methodology to identify abnormal calls is that the proposed method of using non-linguistic features is fairly independent of the language of conversation and does not require the use of a language-dependent speech recognition engine.

As we write this monograph, there are several aspects of call center conversation that are being researched. While one active area is that of building better speech recognition engines by extensive training using deep neural networks [1], there is work being carried out to build language-independent acoustic models to handle conversations that are multilingual. While the dependency on the availability of speech data is not avoidable in all of these cases, there is work being carried out to use minimal amount of data to achieve better ability to convert speech into text. While there exists considerably good speech-to-text conversion for some languages, these are far from acceptable for natural language conversation (office meetings, call center conversation). Given the current performance accuracy of the speech-to-text conversion process, analysis of noisy text is another area of research that has gained importance. From the aspect of emotion, there is a wide range of work identifying the best speech feature set that can effectively assist in exacting emotion of a normal speaker (not acted speech) by analyzing audio. Work in all these areas will significantly benefit our ability to analyze call center conversations.

Reference

1. X. Huang, J. Baker, R. Reddy, A historical perspective of speech recognition. Commun ACM **57**(1), 94–103 (2014)

Appendix A
Informal Definitions

1 *Resource rich language*: A language for which resources are abundant. An example could be that of American English. However, Indian English is not a resource rich language.
2 *Resource deficit language*: A language for which there is no sufficient corpus available. Generally, there is a need for such corpus to train speech recognition engine. In the absence of such corpus, it is difficult to build a speech recognition capability for that language. For example, Hindi.
3 *ASR*: Automatic Speech Recognition is the process of converting audio into transcripts automatically.
4 *Speaker dependent*: A speech recognition system must be speaker independent if the speech recognition system is capable of converting speech spoken by anyone into text with equal ease.
5 *Speaker independent*: A speech recognition system is said to speaker dependent if it works very well for a particular speaker. Such systems are specifically tuned to work well for a specific person. Like dictation systems.
6 *Dictation System*: Is a speaker-dependent system and is specifically tuned to work well for a particular person. Generally, dictation systems require a person to speak a couple of sentences; the spoken data is used to personalize the speech recognition engine.
7 *Lexicon*: A dictionary that captures and maps words to the sounds. This helps the ASR map sounds to human recognizable words. Generally, lexicons are constructed manually by linguists who are familiar with the spoken usage of words.
8 *SLM*: Statistical Language Model is used by ASR to capture the domain-specific spoken language usage.
9 *AM*: Acoustic Models are generally statistical models that model sound or phoneme. This is used by the ASR to convert speech into text. Understandably, one will have different acoustic models for the same language. For instance you would have an AM for Indian English which is different from AM for American English.
10 *PBX*: Private Branch Exchange is a private telephone network used within an enterprise.

© The Author(s) 2015   63
S.K. Kopparapu, *Non-Linguistic Analysis of Call Center Conversations*,
SpringerBriefs in Electrical and Computer Engineering,
DOI 10.1007/978-3-319-00897-4

Appendix B
Computing Speaking Rate

Speaking rate is a measure of the number of spoken words per min (wpm). Measuring speaking rate involves identifying a feature of speech and then calculating the rate by counting the occurrence of that feature per unit time [1]. While there is a general debate on which feature to use, argument persists between choice of syllable and phoneme. Some insist that syllable is the right unit while others oppose that the universal relevancy of syllable is not assessed and that phonemes may be a better candidate. However [2] showed that measuring syllable rate is more correlated to the perceptual speaking rate than measuring the phoneme rate. The correlation is 0.81 using syllable as the feature as against a correlation of 0.73 when phoneme is used as the feature. Pfitzinger [2] further showed that speaking rates calculated in terms of syllable or phoneme for German language have a correlation of 0.6 for normal rate speech. The level of correlation is higher [1] for languages with a simple consonant-vowel syllable structure compared to languages allowing more consonant cluster complexity. Also, at fast speaking rate language-dependent strategies may also influence [3] the computed speaking rate from audio.

We focus on syllable as the measure of speech to compute the speaking rate. The algorithm to detect syllable nuclei in a spoken speech has been described in [4].

The syllable in speech is identified as a function of the intensity of the spoken voice and the voicedness of speech.

A method to identify syllable in spoken speech has been proposed by Nivja and Ton [4]. Local intensity peaks in speech point toward potential syllables since a vowel within a syllable has higher energy than the energy contributed by the surrounding sounds.

The vowel /o/ in the word "WORD" (spelled /word/) has higher energy (intensity) than consonants /w/, /r/, and /d/.

Any peaks that originate in unvoiced regions are discarded to compute the correct number of syllables in spoken speech. Steps involved in speaking rate computation

© The Author(s) 2015   65
S.K. Kopparapu, *Non-Linguistic Analysis of Call Center Conversations*,
SpringerBriefs in Electrical and Computer Engineering,
DOI 10.1007/978-3-319-00897-4

Step 1 Pre processing of the Speech File

Read speech file $(S(t))$
Get the total duration of the speech segment (T)

Step 2 Compute Threshold (τ)

The speech is segmented into blocks of $k = 64\,$ms with an overlap of $l = 16\,$ms to produce $b = \left(\frac{T-k}{l}\right)$ blocks.
Average energy in each of the b speech segment is computed and 0.99 quantile is used to obtain a threshold (τ)

Step 3 Marking pauses and sounding in speech segment

Using τ computed in Step 2, we determine pauses in the speech.
Total pause duration (T_p) in the speech segment is calculated

Identification of pauses in spoken speech enables more accurate computation of the occurrences of the syllables per unit time.

Step 4 Identifying syllables

Mark energy peaks in each frame, these become a possible presence of syllable $(S = \{s_1, s_2, \ldots s_n\})$. The number of syllables is $n = |S|$.

Step 5 Identify valid peaks and discard the invalid peaks

Determine the intensity peaks that are close to one another and retain only one of them. These are valid peaks $(S' \in S$ meaning $|S'| < |S|)$.

Step 6 Identifying Voiced (V) and Unvoiced regions

By computing the pitch (presence of pitch would signify voice) mark all $S'' = \{S' \in V\}$ note that $|S''| < |S'|$

Step 7 Mark syllables Step 4 which occur in voiced regions Step 6.

The speaking rate is computed as $S_{\text{sps}} = \frac{|S''|}{T-T_p}$

Step 8 Computing Speaking rate

we can compute the speaking rate in wpm using a conversion factor of $\gamma = 1.5$ as suggested by Yaruss [5], namely

$$S_{\text{wpm}} = \left(\frac{S_{\text{sps}}}{\gamma}\right) \times 60 \qquad\qquad (B.1)$$

where S_{wpm} is the speaking rate in words per minute, S_{sps} is the number of syllables per second and γ is the conversion factor between syllable and word.

Performance accuracy of the syllable detection was tested on a text "the first one believed in faith, he thought" spoken by nine different people. Note that the number of syllables in this sentence is eight, namely The first one beli.e.ved in faith he

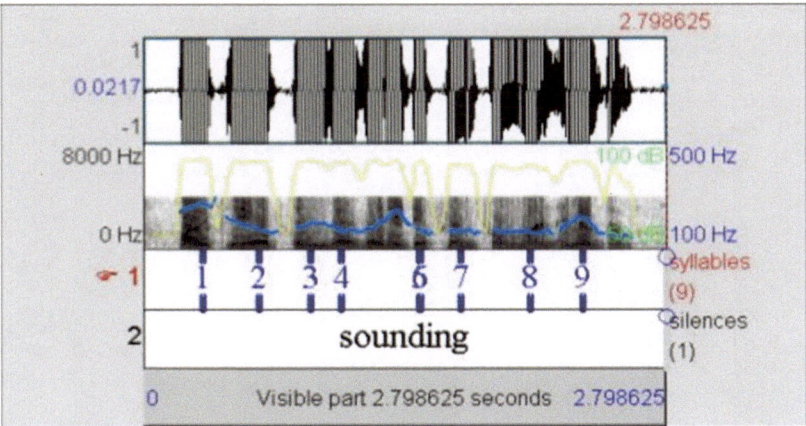

Fig. B.1 The intensity (*yellow line*; the peaks show the presence of a syllable) and pitch (*dark blue* in spectrum; used to identify voiced and unvoiced speech). Syllables identification for the sentence /*all three of them were very different from*/ which has nine syllables

thought. Then using the procedure described earlier we computed the number of syllables from the spoken audio (by nine different users) and found them to be eight in each of the spoken case. This demonstrates the ability to extract the right number of syllables from speech. In another experiment, a short paragraph, spoken in English by 10 different people with three variations in speaking rate was recorded. The number of syllables identified using the algorithm based on syllable detection for these 30 spoken speech was within ±10 % of the actual number of syllables, present in the paragraph text [5]. Figure B.1 shows an example of syllable identification in a spoken English sentence.

Appendix C
Estimating P_{**}

Let us assume that the call conversation has been marked into segments that have been spoken by agent (AGENT SPEAK), by customer (CUSTOMER SPEAK) or was on hold (HOLD MUSIC).

Figure C.1 shows the method used to compute all the probabilities P_{**} and the size of the node for a call conversation of 23 s duration. In terms of notation, the segments marked in black (dark) represent the portion of the call where the agent is speaking (AGENT SPEAK), segments marked in white represent the portion of the call where customer is speaking (CUSTOMER SPEAK), and the segments marked in blue represent the call where the customer is on hold (HOLD MUSIC).

The probability P_{ac} is the probability of a customer speaking after the agent has spoken, P_{ca} is the probability of agent speaking after the customer has spoken, P_{am} is the probability of the agent putting the customer on HOLD MUSIC and P_{ma} is the probability of the customer conversing with the agent after being on hold.

The directed graph model represents the overall picture of the complete call. The size of the node AGENT SPEAK represents the total duration of the conversation when the agent was speaking (12 s in Fig. C.1), similarly the size of the node CUSTOMER SPEAK captures the total duration for which the customer was speaking during the entire call (9 s), and the size of the node HOLD MUSIC represents the total duration for which the customer was on hold (2 s). Clearly, in this sample call it can be visualized (from the size of the node) that the agent has spoken for the longest duration.

Similarly we can compute, automatically, the various probabilities P_{**} as follows. As seen in Fig. C.1 there are a total of six instances when the customer spoke after the agent (marked by P_{ac} in Fig. C.1) and only one instance of the music coming into existence after the agent spoke (marked as P_{am}). So, we can compute $P_{ac} = \frac{6}{7}$ and $P_{am} = \frac{1}{7}$. On the other hand, as seen in Fig. C.1 there are six transitions from CUSTOMER SPEAK to AGENT SPEAK and no transition from CUSTOMER SPEAK to HOLD MUSIC, so we compute $P_{ca} = \frac{6}{6}$ and $P_{cm} = \frac{0}{6}$. Similarly, there is only one

© The Author(s) 2015
S.K. Kopparapu, *Non-Linguistic Analysis of Call Center Conversations*,
SpringerBriefs in Electrical and Computer Engineering,
DOI 10.1007/978-3-319-00897-4

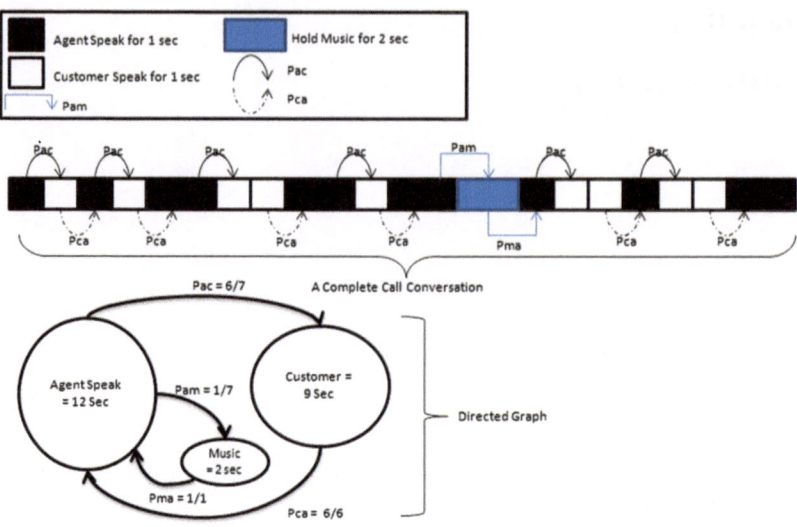

Fig. C.1 Computing P_{**} from a complete call

transition from HOLD MUSIC to AGENT SPEAK and no transition from HOLD MUSIC to CUSTOMER SPEAK. Using this we can compute $P_{ma} = \frac{1}{1}$ and $P_{mc} = \frac{0}{1}$.

Appendix D
Best Sample Size for Computing Real-Time Speaking Rate

To determine the smallest sample size of the audio to compute the speaking rate, experiments were carried out on two agent-customer audio call conversations. At first we calculate the speaking rate (SR0) of the complete call (Level 1 in Fig. D.1) and then divide it into halves (Level 2 in Fig. D.1) and for each of these halves we compute the speaking rate SR01 and SR02 separately. Figure D.1 shows the division of the speech file into different multiresolution levels and the corresponding computed speaking rate.

Let SR0; (SR01, SR02); (SR011, SR012, SR021, SR022)); (SR0111, SR0112, SR0121, SR0122, SR0211, SR0212, SR0221, SR0222) be the computed speaking rate at Level 1, Level 2, Level 3, and Level 4 respectively. An automated method to split the audio call conversation at each level was adopted and the speaking rate computed for each audio segment. We further compute the mean (μ) and variance (σ^2) of the speaking rate for all the speech files at the same level. For example, at Level N, there would be $\log_2 N$ distinct speech files and hence $\log_2 N$ speaking rates computed. The (μ, σ^2) of these $\log_2 N$ speaking rates were computed. A boxplot was used to represent the speaking rate at each level.

Figures D.2 and D.3 shows the boxplot of two different sample call conversations of duration 272.8 and 150.4 s respectively, for different levels.

As can be seen in Fig. D.2, the computed speaking rate until Level 7 is fairly close to the speaking rate of the entire speech file (Level 1); however, the speaking rate sees a large deviation for Level 8 and Level 9. The length of the speech files used to compute the speaking rate at Level 7 is 4.26 ($\approx$5) s.

Similarly, Fig. D.3, shows that the speaking rate computing detoriates for Level 7 and upward. As can be seen the speaking rate computed at Level 7 captures the actual speaking rate of the speech sample, suggesting that the smallest duration of speech sample that can be used to compute the speaking rate reliably is 4.7 ($\approx$5) s.

We conclude that a 5 s speech file is the smallest duration of the speech file that can be used to reliably compute the speaking rate.

© The Author(s) 2015
S.K. Kopparapu, *Non-Linguistic Analysis of Call Center Conversations*,
SpringerBriefs in Electrical and Computer Engineering,
DOI 10.1007/978-3-319-00897-4

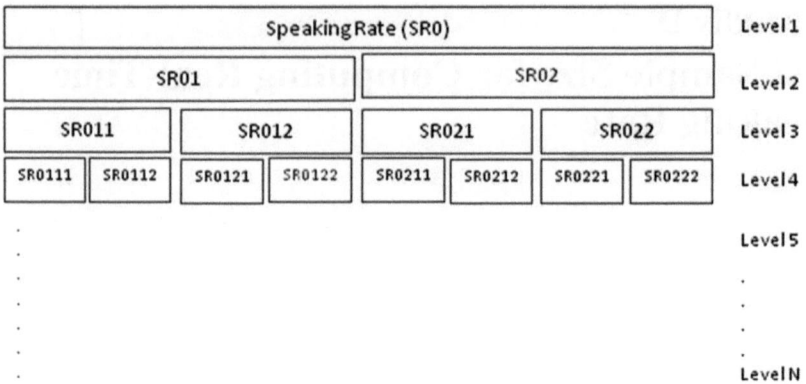

Fig. D.1 Speaking rate computing at different resolutions

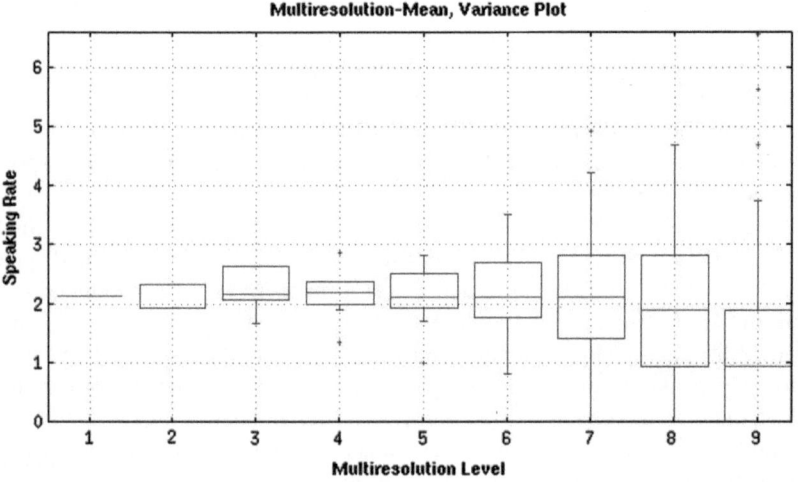

Fig. D.2 Box plot of speaking rate at different resolutions for a sample call conversation (272.8 s)

A boxplot (see Fig. D.4) is a graphical display of five number summary. The four steps followed in plotting a boxplot are

1. *Draw a box from the 25th to the 75th percentile.*
2. *Split the box with a line at the median.*
3. *Draw a thin line (whisker) from the 75th percentile up to the maximum value.*
4. *Draw another thin line from the 25th percentile down to the minimum value.*

The length of the box in a box plot, namely the distance between the 25th and the 75th percentiles, is known as the interquartile range (IQR).

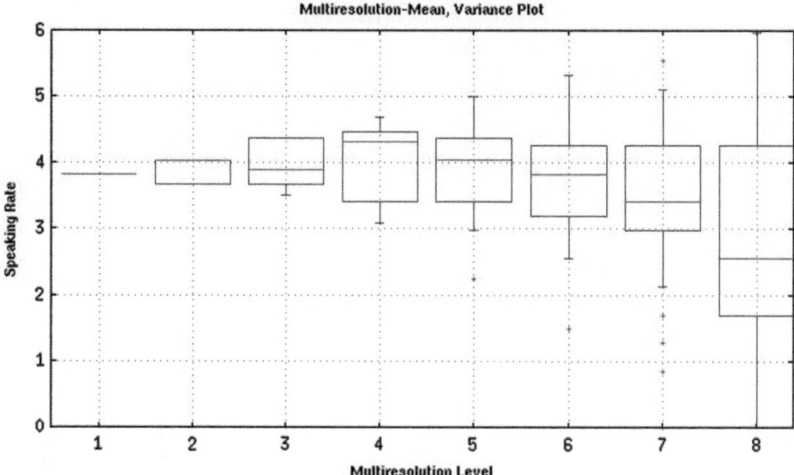

Fig. D.3 Box plot of speaking rate at different resolutions for a sample call conversation (150.4 s)

Fig. D.4 A sample box plot

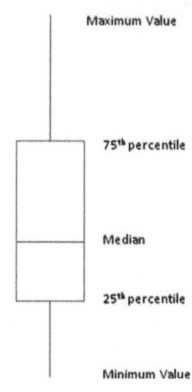

Appendix E
Resource Deficient Language ASR

Automatic Speech Recognition (ASR) technologies have primarily focused on a handful of languages and subsequently these languages have access to language resources such as annotated corpora and pronunciation dictionaries.

In practice a speech corpus is a must for building acoustic models (AM) for speech recognition, while building language models (LM) or grammars for speech recognition may require only a text corpus. Language models and grammars are application and domain specific while AMs are independent of domain. Previous work on language models for speech recognition of resource-deficient languages [6–10] has discussed building of language models using machine translation of text in resource rich languages. However, availability of a speech corpus for a specific language has been an essential requirement to build acoustic models and speech recognition-based solutions thereof in the respective language. A typical speech corpus is a set of audio files and its associated transcriptions.

The process of creating a speech corpus in any language is a laborious, expensive, and time-consuming process, which means several languages do not have a speech corpus available; especially when the language has no viable commercial speech recognition-based solution. Thus there exists a long-felt need for an inexpensive speech corpus.

In a multilingual country like India, where there are 22 officially recognized distinct languages, a speech solution has to work in different languages to truly address a large multilingual population. Further, most of these languages are 'resource-deficient' in terms of availability of the language resources required for building speech recognition capability.

The process of creating a speech corpus in any language is a laborious, expensive, and time-consuming process. The usual process of speech corpus creation starts with a linguist determining the language-specific idiosyncrasies and then a textual corpus is built to take care of the even distribution of the phonemes in the language (also called phonetically balanced corpus). Subsequently a target speaker age, accent, and gender distribution is computed leading to the recruitment phase where participants or the speakers are recruited. The actual speech recording is then undertaken from the recruited speakers in predetermined environments. Typically, the text corpus is

© The Author(s) 2015
S.K. Kopparapu, *Non-Linguistic Analysis of Call Center Conversations*,
SpringerBriefs in Electrical and Computer Engineering,
DOI 10.1007/978-3-319-00897-4

created by keeping the underlying domain in mind for which the speech recognition is going to be used. For spontaneous conversational speech like Telephone calls and Meetings, the process of speech corpus creation may start directly from the speaker recruitment phase. Once the speech data is collected, the speech is carefully heard by a human who is a native speaker of the said language and transcribed manually. The complete set of the speech data and the corresponding transcription together forms the speech corpus. This is quite an elaborate process, which means several languages do not have a speech corpus available, especially when the languages do not have commercial speech recognition-based solution viability. Thus there exists a long-felt need for an effortless and inexpensive method and system that enables creation of a speech corpus.

Exploiting existing collections of online speech data to build an inexpensive and usable speech corpus is a way to build speech copra for resource deficit languages. In [11] we proposed a frugal method for speech corpus creation using existing speech data available on the Internet. Speech data is available on the web, in the form of news, audio books, video talks, lectures, etc., and is accompanied by the transcripts. Such data are often available in different languages. For instance, the All India Radio (AIR) website [12] provides archives of news in various Indian languages. In a way one has access to a well-transcribed speech data, albeit with certain constraints in terms of (a) limited speaker variability (number of speakers), (b) limited environment (recording environment), and (c) limited domain. A combination of such freely available speech corpus and a smaller amount of focused traditionally collected speech data can enable construction of a speech corpus for a resource deficient language [11].

> Construction of speech corpus is frugal in the sense, it is less expensive, less laborious, and less time-consuming to construct the speech corpus. Frugal speech corpus constructed in such a manner can be used for training acoustic models for a resource deficient language, for use in ASR.

Let us say we need to create a speech corpus for a language L. Identify web-pages which have public access to speech data of this language. An automatic process downloads the speech data and the corresponding transcription. Automatic speech alignment algorithms [13–15] match the transcription to the speech file. Now analyze the transcripts using language processing [16] to identify those text segments that would satisfy the phonetic balancing of the speech corpus. A large portion, say X % of the speech corresponding to the collected text can come from the speech data from the Internet itself and the remaining $(100 - X)$ % could be collected in the usual way. The choice of X would determine the amount of effort, time, and expenditure in constructing the speech corpus. The larger the X the more frugal the construction of the speech corpus. In the limiting case, when $X = 0$, it would be what is conventionally used for speech corpus creation. On the other extreme, $X = 100$ one would have access to the cheapest mode of creation of speech corpus at the cost of lack of diversity in terms of speaker variability, environment. If one could control X based on where one would like to use the speech corpus for speech recognition engine training.

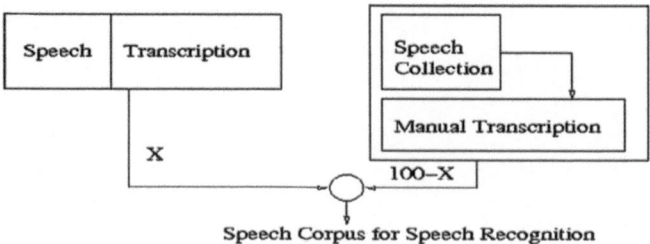

Fig. E.1 Block diagram of the frugal speech corpus creation

Figure E.1 depicts the suggested approach. The left-hand side shows the use of already available resources on the web and the right-hand side shows the conventional speech data collection process. The fact X determines the amount of deviation from the conventional approach in terms of making it frugal.

Appendix F
WER Conversational Speech

Though the literature claims to give a WER of less than 25–30 % for conversational speech, our own finding has been very different. We have analyzed 6, 000 realistic call center conversations in an insurance firm, which were of approximately 5 min duration on average.

We manually transcribed about 66 calls and used about 40 of these conversational transcripts to build a n-gram (n=3) statistical language model. Understandably, there were several proper names, to cater to this a support lexicon was manually created. We used Sphinx with default AM (WSJ 8 KHz) because the conversations that we were looking at were in American English.

We observed a WER of 40–50 % on a dataset of 66 calls. The best accuracy in terms of WER was 30 % and the worst was as high as 75 %. On average the WER for 20 randomly selected call conversation from the set of 6000 calls was >50 %.

References

1. P. Francois, C. Christophe, M. Egidio, Across-language perspective on speech information rate. Language **87**, 539–558 (2011)
2. H. Pfitzinger, Local speaking rate as a combination of syllable and phone rate, in Proceeding of ICSLP, (1998).
3. F. Ramus, Acoustic correlates of linguistic rhythm: perspectives, in Proceeding of International Conference on Speech Prosody, (2002).
4. N.H. De Jong, T. Wempe, Praat script to detect syllable nuclei and measure speech rate automatically. Behav. Res. Methods **41**, 385–390 (2009)
5. J. Scott, Yaruss, Converting between word and syllable counts in children's conversational speech samples. J. Fluen. Disorders **25**(4), 305–316 (2000)
6. T. J. Arnar, W. D. Edward, I. Koji, S. Furui, Language model adaptation for resource deficient languages using translated data, in INTERSPEECH'05, pp 1329–1332, (2005).
7. T.J. Arnar, W. Edward, I. Koji, S. Furui, APSIPA Annual Summit and Conference Sapporo, Japan, (2009).
8. W. Kim, S. Khudanpur, *Language model adaptation using cross-lingual information, in Proceedings of* (Eurospeech Switzerland, Geneva, 2003), pp. 3129–3132

© The Author(s) 2015
S.K. Kopparapu, *Non-Linguistic Analysis of Call Center Conversations*,
SpringerBriefs in Electrical and Computer Engineering,
DOI 10.1007/978-3-319-00897-4

9. I. Dawa, Y. Sagisaka, S. Nakamura, Investigation of asr systems for resource-deficient languages. ACTA AUTOMATICA SINICA **1**, 1–8 (2008)
10. S.K. Kopparapu, I.A. Sheikh, Enabling rapid prototyping of an existing speech solution into another language in Proceedings of Oriental COCOSDA, Hsinchu, Taiwan, (2011).
11. I. A. Sheikh, S. K. Kopparapu, A frugal method and system for creating speech corpus. Indian Patent Application 2148/MUM/2011 (2011).
12. AIR. All india radio news archives.
13. Y. Tao, L. Xueqing, W. Bian, A dynamic alignment algorithm for imperfect speech and transcript. Comput. Sci. Inf. Systems **7**, 75–84 (2010)
14. A. Katsamanis, M. Black, P. G. Georgiou, L. Goldstein, S. S. N. Sailalign, Robust long speech-text alignment, in Proceedings of Workshop on New Tools and Methods for Very Large Scale Research in Phonetic Sciences, pp 28–31, Pennsylvania, (2011).
15. I. Ahmed, S.K. Kopparapu. Technique for automatic sentence level alignment of long speech and transcripts, in F. Bimbot, C. Cerisara, C. Fougeron, G. Gravier, L. Lamel, F. Pellegrino, P. Perrier (eds), INTERSPEECH, pp 1516–1519. ISCA, (2013).
16. M. Liang, R. Lyu, Y. Chiang, An efficient algorithm to select phonetically balanced scripts for constructing a speech corpus, in Proceedings of Workshop on New Tools and Methods for Very Large Scale Research in Phonetic Sciences, pp 433–437, Beijing, (2003).

Index

© The Author(s) 2015
S.K. Kopparapu, *Non-Linguistic Analysis of Call Center Conversations*,
SpringerBriefs in Electrical and Computer Engineering,
DOI 10.1007/978-3-319-00897-4